Causation: The Basics explores questions about what causes are, and how we come to know them, describe them, and put them to use. The book begins with an introduction to the history of philosophical thinking about causation, followed by a series of chapters introducing important contemporary accounts of causation. It concludes with chapters on causation and agency, causal discovery, and causal explanation. Key questions explored in the book include:

- What distinguishes correlation from causation?
- How are the causes of singular events related to more general patterns of cause and effect?
- How are commonsense, scientific, and legal conceptions of causation related?
- Can certain occurrences be singled out as the main or principle causes of some effect?
- Is there a place in the world's causal structure for human agency and free will?

While introducing the major philosophical debates about the nature of causation, *Causation: The Basics* emphasizes the uses and challenges of causal reasoning as it occurs in the sciences, engineering, medicine, and other areas of human life. With a glossary of key terms and suggestions for further reading, the book provides readers with a clear and concise introduction to both theoretical and practical questions about causation.

Stuart Glennan is the Harry Ice Professor of Philosophy at Butler University. He has published widely on causation and explanation in philosophy of science and is the author of *The New Mechanical Philosophy* (2017) and co-editor with Phyllis Illari of the *Routledge Handbook of Mechanisms and Mechanical Philosophy* (2018).

The Basics Series

The Basics is a highly successful series of accessible guidebooks which provide an overview of the fundamental principles of a subject area in a jargon-free and undaunting format.

Intended for students approaching a subject for the first time, the books both introduce the essentials of a subject and provide an ideal springboard for further study. With over 50 titles spanning subjects from artificial intelligence (AI) to women's studies, *The Basics* are an ideal starting point for students seeking to understand a subject area.

Each text comes with recommendations for further study and gradually introduces the complexities and nuances within a subject.

Business Anthropology
Timothy de Waal Malefyt

Eating Disorders
Elizabeth McNaught, Janet Treasure, and Jess Griffiths

Truth
Jc Beall and Ben Middleton

Perception
Bence Nanay

C.G. Jung's Collected Works
Ann Yeoman and Kevin Lu

Corporate Finance
Terence C.M. Tse

Film Genre
Barry Keith Grant

Religions and Sports
Terry D. Shoemaker

Critical Theory
Martin Shuster

For a full list of titles in this series, please visit www.routledge.com/The-Basics/book-series/B

CAUSATION

THE BASICS

Stuart Glennan

Routledge
Taylor & Francis Group
NEW YORK AND LONDON

Designed cover image: © Unsplash

First published 2024
by Routledge
605 Third Avenue, New York, NY 10158

and by Routledge
4 Park Square, Milton Park, Abingdon, Oxon, OX14 4RN

Routledge is an imprint of the Taylor & Francis Group, an informa business

ISBN: 978-1-032-06153-5 (hbk)
ISBN: 978-1-032-06154-2 (pbk)
ISBN: 978-1-003-20095-6 (ebk)

DOI: 10.4324/9781003200956

Typeset in Bembo
by Apex CoVantage, LLC

For
Paul, John, and Elliot,
who have caused me such joy

CONTENTS

PREFACE

It turns out to be a challenge to write a book about the basics of causation. While I've spent a good bit of my career thinking about the topic, deciding what is basic to an understanding of causation required me to step away from my particular concerns and think about the question of what anyone should know, and what can be made accessible to a non-specialist audience. This is a problem for any book about the basics, but the problem feels especially acute for causation, because it is touched on in so many disciplines and professions—philosophy, but also the natural and social sciences, history, statistics, engineering, the law, and public health, to name a few. I have tried to strike a balance between the conceptual analysis of philosophers and the empirical methods of scientists, engineers, and others who engage in causal discovery and reasoning, and whenever possible to put these different approaches to thinking about causation into dialog. I have also tried, as much as possible in a short introductory book, to avoid toy examples, and to illustrate approaches by considering how they illuminate complex problems about causes, from income inequality to climate change.

By design this book is short, minimizes technical vocabulary, limits references, and has no footnotes. Inevitably there are many topics I've had to ignore, many details I've had to gloss over, and many times where I haven't been able to acknowledge all of the important contributions of scholars past and present to our understanding of causation. These were painful but necessary decisions, and I hope that readers who know the field will find them reasonable.

I'd like to thank Andy Beck, Marc Stratton, and the team at Routledge for their support throughout this project. I am grateful to Andy for suggesting I write this book—something I'd never considered but instantly wanted to do. Special thanks are due to Kristian González Barman, Olaf Dekkers, and Harold Kincaid, who read the entire manuscript and offered valuable comments. I won't attempt to name my many teachers, friends, and interlocutors whose thoughts have shaped this book, but I do want to express my gratitude and the hope that I've done their ideas justice. Finally, I must thank my wife Lesley, whose patience and encouragement has given me the opportunity and will to complete this project.

1

SOME BASIC QUESTIONS ABOUT CAUSATION

Choose anything you like in the world—any event, fact, pattern, or behavior; any object, artifact, or living thing—and you can ask what caused it to be. Causal thinking is a very basic part of our cognitive make-up, acquired in infancy, which allows us to navigate our world. As infants we learn that our actions make things happen: dropping the cup causes it to spill, and smiling or crying causes reactions from our parents. And as our cognitive capacities develop, we come to understand more of the world's causal structure. We begin to recognize that there are chains or webs of causes, with one event causing others, which cause others in turn. We come to understand regularities in the world's causal structure—that certain kinds of events typically or invariably bring about others, and so on.

Identifying causes is also a central occupation of natural science. Given any phenomenon in nature—changing tides, global warming, the color of our eyes or skin, or a new and deadly virus—we look to scientists to explain what caused or causes these things to be. Social scientists seek causal explanations of phenomena like poverty, economic booms and busts, electoral outcomes, racism, violence, and health care disparities. Engineers and technologists need causes to understand how they can build or repair things—from cars to bridges to computers. Understanding causes is important to the law, since in the law we want to hold to account agents whose actions caused harms. It is essential to

DOI: 10.4324/9781003200956-1

medicine and public health, because without a knowledge of causes we will be limited in our ability to treat or prevent disease. Understanding causes is also important to politics and policy, because in these domains we seek to act in ways that will cause the outcomes we desire.

Although causal talk and reasoning is everywhere both in everyday life and science, as we look more closely, we will discover that it is often difficult to say what causes what. Sometimes our problem is that we lack knowledge. Perhaps we have become infected by a virus, but we don't know where and from whom we caught it. But other problems are more conceptual and general. Just what kind of things should count as causes? Many things we think of as causes are events—happenings at particular places and times—but can things that don't happen cause things as well? Is my failure to brake at the stop sign a cause of the accident? Also, can we speak of persons or objects as causes (or effects)? Can we say for instance that the accident was caused by the construction barrier, or that the toddler caused the spill? Do beliefs or emotions or dispositions count as causes? Also, given that a close inspection of any happening in the world shows that it depends upon many things, should we try to single out only some of these as causes (or even *the* causes) of an effect? Surely, we should count a lightning strike as a cause of a forest fire, but should we also count the drought which made the forest so flammable, or even the oxygen without which no fire would occur? And how should we understand the relationship between the causes of a specific event like a particular forest fire and more general causal knowledge about what causes forest fires?

This book is about the basics of causation. Its aim is to give the reader tools to understand how the world is causally structured, to assess claims about what causes what, and to notice causes that may have been hidden from view. In it I bring together two different approaches that are sometimes in tension. The first half of the book largely focuses on philosophical approaches that analyze the concept of cause, while the second half is more concerned with the scientific practices which provide us with causal knowledge, as well as the concrete ways in which scientists, engineers, lawyers and policy makers put causes to use. While these approaches are sometimes in tension, my own view is that both are necessary to help us understand what causation is and does.

The topics and theories related to causation are varied, but there are some common questions and themes that run through them. To help you navigate this landscape, I want to lay out in this introduction a few key questions that dominate philosophical and scientific inquiries into causation. Noticing how different thinkers have addressed these questions can clarify how competing approaches to causation are related, and make clear to what domains the tools and theories they introduce may be applied.

WHAT KIND OF QUESTION ABOUT CAUSATION?

Our first question is about the kinds of questions we need an account of causation to answer. Most such questions can be sorted into three categories: conceptual/semantic, metaphysical, or epistemological/methodological. Conceptual and semantic questions are often the starting point for philosophical analysis, and it is arguably through such analysis that philosophy provides its most distinctive contribution to understanding causation. That analysis will help us understand the meaning of and relation between various causal concepts, and describe what it takes for various kinds of causal statements to be true.

In contrast to conceptual and semantic questions, which are about language and meaning, metaphysical (or ontological) questions are questions about the nature of reality independently of whether or how we think of it. When we ask what is the nature of causation, or whether everything has a cause, we are asking metaphysical questions.

Conceptual and metaphysical questions are not unrelated, since it is only through language and concepts that we can make claims about what is in the world. Consider this apparently simple question: What is water? The answer that might immediately come to mind is a scientific one. It is a molecule, H_2O. But this is hardly all there is to water. Humans had a concept of water long before they understood its chemical composition. It is something that fills lakes and streams and falls sometimes from the sky, and it is something that slakes our thirst, and that sustains plants and animals. Concepts of water may vary depending upon time and culture. What water is to indigenous peoples of Central America might not be quite what water is to ancient Greeks or contemporary Norwegians.

You might think that the conceptual questions are easily separated from the scientific questions, and that science can tell us what, from an ontological point of view, water really is. The problem though is that science cannot describe "the facts" without concepts: when we describe water as H_2O, we are using chemical concepts—elements, molecules, bonds, and so forth. But these concepts depend not just upon how the world is but also upon our cognitive and linguistic abilities, and our current state of knowledge. As we come to better understand the world, we often need to revise our concepts.

To many the word "metaphysics" has mystical connotations, suggesting what you may think of as irresolvable questions about whether there is a God, whether we have free will, or whether the universe we inhabit is all there is. Metaphysics, you might think, if it has any sense at all, is a realm of speculation that lies beyond science. But on the view I shall adopt, there is no bright line between the scientific and the metaphysical. Empirical knowledge should inform our views on metaphysical questions, like whether and in what sense we have free will, but answers to such questions cannot simply be read off the results of scientific observation and experiment. Answers to metaphysical questions are more abstract, and they offer different ways of interpreting the world we find ourselves in. Questions about causation are metaphysical in this sense. We can ask science to identify the causes of particular things, but the idea that the world has causal structure seems to be more basic—perhaps something built into our cognitive apparatus or something we must assume in order to do science at all.

The third category of questions we may ask about causation are epistemological or methodological ones—questions about what we can know about causes, how we can find them, and what uses we can put our causal knowledge to. Here again we can look to science, but we must be aware that scientific practice often makes epistemological assumptions about causation. Philosophical analysis can help identify these assumptions, so that we can understand why and when they are justified.

The most basic epistemological question about causation concerns how we come to know causes. Can we observe causes in the world directly, or must we infer them? And what roles do experiments play in causal discovery and inference? Another question concerns how to use models and statistical techniques to represent and identify causal relationships. A third important question concerns how we use causes to explain why things happen, and why we pick out certain things as causes while leaving other things out. Why, for instance, do we say a forest fire was caused by a lightning strike, but not by the wood that burned, or even the oxygen needed for combustion? And why or when might it be better to say that the fire's cause was forest management policies that increased fire risk?

As we explore accounts of causation in this book, you will find a constant back and forth between conceptual, metaphysical and epistemological/methodological questions. This is because a proper understanding of causation will require answers to these questions to hang together. We must understand how causal concepts and models are used to characterize the world's causal structure, and our account of that structure needs to make sense of how, why, and when we can know causes. In comparing the theories we'll discuss, you'll want to consider which kinds of questions motivate them, and which questions they leave unanswered.

SCIENCE OR COMMON SENSE?

A second key question in our analysis of causation concerns whether causal concepts are best understood as commonsense concepts (sometimes called folk concepts) or as scientific concepts. As I noted at the outset, causal thinking is something humans engage in from childhood, and which is essential for making our way through the world. Commonsense causal reasoning helps us understand happenings in the world, from a flood in the basement to last year's election. It can explain and predict the behavior of our friends, children, lovers and strangers, and also to help achieve our goals, from making lunch to getting into college. On the other hand, the sciences have developed special tools for causal discovery and inference, including techniques for conducting controlled experiments and algorithms for identifying causal patterns in statistical data.

There is actually some dispute among scientists about whether causes should be part of our scientific account of nature at all. The philosopher Bertrand Russell famously characterized causality as "a relic of a bygone age, surviving, like the monarchy, only because it is erroneously supposed to do no harm" (1913 p. 1). The history of statistics also shows considerable anxiety about causation and causal reasoning. Although statistical techniques can identify correlations and other predictive relationships between variables, the fact that variables are correlated is not enough to tell you which variable is the cause and which the effect, or whether they are both effects of some common cause that accounts for the correlation.

Philosophers sometimes suggest that scientific and commonsense concepts of causation are different (Hitchcock 2007). Scientists are generally concerned with uncovering pieces of the world's causal structure, understanding how various underlying causal relationships and processes will allow us to predict, explain and intervene in the world. In commonsense contexts, people are often more concerned with picking out certain objects, individuals or events as THE causes of events, often to assess responsibility or blame. Suppose for example that I turned on the tap in the tub, walked away, forgetting to turn it off, and later discovered my basement flooded. I would likely identify my forgetting as the cause of the flood. I have only myself to blame! But one might insist from a scientific or metaphysical point of view that things I didn't do can't cause things to happen. What caused the flood, was the water flowing and nothing more. And were I to insist that it was my failure to turn off the faucet, you could reasonably ask why I am picking on me, when my wife could also have turned off the faucet.

There are, however, lots of things that scientific and commonsense reasoning about causes have in common. The kind of causal reasoning used by an experimental physicist in locating a malfunction in her apparatus is not so different from the kind of causal reasoning I engage in when I figure out why my basement is flooded. Moreover, there are many contexts, for instance in law and politics, where commonsense concerns with assigning blame and responsibility interact with scientifically informed accounts of the world's causal structure. As we consider differing accounts of causation, we should watch how well they make sense both of commonsense and scientific cases.

SINGULAR OR GENERIC?

Causal claims come in two basic varieties, as illustrated by the following sentences:

- Smoking causes lung cancer
- George's smoking caused him to get lung cancer

The first sentence asserts a causal relationship between two kinds of activities and events. I'll refer to sentences like these as generic causal claims; they are also called general or type causal claims. The second sentence asserts a causal relationship between two particular activities or events. I'll refer to these as singular causal claims; they are also called particular, actual, or token causal claims.

What is the relationship between these claims? You might think that the generic claim is what makes the singular claim true: George's smoking caused his lung cancer, because George smoked, and smoking causes lung cancer. But this reasoning isn't necessarily right. Not everyone who smokes gets lung cancer, and some who don't smoke do get lung cancer. Moreover, it seems entirely possible that even if George smoked and got lung cancer, that it wasn't his smoking that caused his lung cancer. Perhaps then the relation between singular and generic claims goes the other way: the generic causal statement is true because some set of singular causal statements is true. But if this is how the dependence runs, what explains the general pattern?

Figuring out the relationship between singular and generic causal claims has been a central concern of philosophical inquiries into causation, and the question comes up at many places in the book. As you work through the chapters, you will find it helpful to attend to which kinds of causal claims are under discussion, and how different approaches to causation see them as related. It is also important to notice the different varieties of generic causal claims. Some generic claims imply that the cause always brings about the effect, while others imply only that it sometimes brings about the effect, and still others, like the claim that smoking causes cancer, seem to be claims about making the effect more likely. What generic causal claims have in common is that they describe patterns of causal dependence, but the patterns are of different kinds.

CONNECTION OR DIFFERENCE MAKING?

Philosophical accounts of causation have to make sense of two features that we generally recognize as characteristic of the relation between causes and effects. One is that causes are connected to their effects, and the other is that causes make a difference to their effects. The idea of connection is roughly this: Whenever a cause is followed by an effect, there needs to be something—call it a connecting causal process—by which the cause *produces* the effect. For instance, if we flip a switch, and a light goes on, we assume that for the switch to have actually caused the light to go on, there must be some process by which switching the switch directs current through the fixture to produce the light. We may know little or nothing about what the connecting process is, but we believe it must be there. Another way of putting this idea is that whenever something causes another thing, there must be a mechanism by which the causing occurs.

What does it mean to say that causes make a difference to their effects? We often express the difference a cause makes in counterfactual terms. We say that flipping the switch caused the light to turn on because, *if I had not flipped the switch, then the light would not have turned on.* For generic causal claims we can express difference-making in terms of patterns of association between the cause and effect. In the case of smoking and lung cancer, the difference is probabilistic. Smoking makes lung cancer more likely. At other times, the difference is that the cause is always followed by the effect. The generic causal claim that lowering the temperature of a volume of water to −10° C causes it to freeze implies that whenever you lower the temperature of a volume of water to −10° C it will freeze.

In most causal relationships you will find both connection and difference-making, but occasionally you can get one without the other. The case of my forgetting to turn off the water is a case where you get difference-making without connection. My forgetting certainly made a difference to the flooding, because, had I not forgotten there would have been no flood. Still, it doesn't seem possible that my forgetting produced the flood or was connected to the flood. How could something that didn't happen produce anything at all?

Connection without difference-making occurs in cases of overdetermination, where there are multiple causes each sufficient for an effect. A classic case is the firing squad. If three soldiers each fire a shot at the prisoner's heart, no single shot makes a difference to the prisoner's death, because, had the first soldier missed or failed to shoot, the prisoner still would have died. Nonetheless, there is a causal process connecting the soldiers' shots to the prisoner's death.

Philosophical theories of causation can generally be classified as connection theories (also called production theories) or difference-making theories, depending upon which criterion they take to be fundamental, and these different theories will offer varying accounts of what connection or difference-making amounts to.

REDUCIBLE OR FUNDAMENTAL?

A central strategy in both science and philosophy is reduction. Put roughly, to reduce *X* to *Y* is to show that *X* is somehow made up of, or depends upon, or is defined in terms of *Y*. Take for example, genes and DNA. When scientists discovered that chromosomes were strands of DNA, and that the factors we called genes were identifiable as segments within these strands, they in some sense reduced genes to DNA. While there is no perfect and general way to define genes in molecular terms, genes depend upon DNA, and in some sense genes are "nothing but" bits of DNA (or sometimes RNA). Reduction can be iterated. DNA is a complex macro molecule, and we can explain what it is by looking at its structure. DNA, we might say, is nothing but a double helix bound together by sequences of bases (A, C, G, T). We can continue reducing, but at some point we will reach a point where we arrive at something that we believe to be fundamental, which cannot be reduced to something else. Most contemporary physicists think that what is fundamental is the set of subatomic particles described by what is called the standard model of particle physics. These fundamental entities cannot be reduced any further, and all the standard model can say of them is how they are related to each other.

Reduction in philosophy follows a similar path, though it is generally a conceptual reduction, which aims to show how one concept depends upon or can be defined in terms of others, rather than how one kind of entity depends upon other kinds of entities from which it is made. In the coming chapters we will consider several attempts to reduce causes to something else. The first, and for many years most important, is David Hume's attempt to reduce causal connection to regularities (constant conjunctions). On Hume's view, what it means to say that *A* causes *B* is that all *A*s are regularly followed by *B*s. Another prominent theory attempts to reduce causal dependence to counterfactual dependence. On counterfactual accounts, to say that *B* causally depends upon *A* just means that if *A* had not occurred, *B* would not have occurred. Attempts to reduce causation are generally motivated by concerns about clarity and simplicity. It is thought that what causation is or how we can know it is not well understood, and that by reducing it to something that we understand better, like observations of constant conjunction, we get clearer on these issues.

Reductive approaches to causation can be resisted for two reasons. First, causation may be a fundamental feature of the world, which we use to explain other things, but which we cannot reduce or explain away. For instance, causes may be needed to explain regularities or counterfactual dependence, rather than being reducible to them. Second, we might think that even if a reduction is possible, that any reduction will miss important things about how causes work. A common complaint about reductionism in the sciences is that it can miss the forest for the trees. For instance, while it is true in a sense that genes are nothing but bits of DNA, that kind of reduction can miss the functional role of genes, the information they carry, and the ways in which gene expression is modified by genetic, cellular and larger contexts. We might similarly wonder if a reductive account of causation fails to inform us about important questions, like how we can recognize and use causes in the world.

FACT OR CONVENTION?

A last question to ask about causation concerns whether what the causes of an effect are is simply a matter of fact, or whether, alternatively, our counting things as causes depends in part upon convention. Many causal claims appear to be true simply in virtue of facts about the world. It is a straightforward fact that, all things being equal, saltwater freezes at a lower temperature than fresh water. Other claims though depend upon social or legal conventions. In most states within the United States, children become adults on their 18th birthday. This fact has important social and legal consequences, but it does not derive from any naturally detectable change that happens on one's eighteenth birthday. The biological, psychological and social traits of adulthood appear gradually and at different rates in different individuals, so any natural boundary between child and adult will be vague.

As we explore causation, we will have to figure out which facts about causation are more like the facts about freezing point, and which facts are more like the facts about becoming an adult. Consider this question: Did your grandmother's birth cause her death? Many would be inclined to say no, but the birth clearly made a difference to her dying. Had your grandmother not been born, she wouldn't have died. Still most would be inclined to identify something more immediate and life-threatening, like a stroke or heart attack, as the cause of her death. Our challenge will be to identify the principles by which we select causes from among the many events upon which a given effect depends, and to explore how these principles depend upon facts about the natural world, and how they depend upon our practical—social, legal, moral, etc.—goals and interests.

LOOKING FORWARD

Here is a brief summary of what is to come. We'll begin our investigation in Chapter 2 with a look back into the history of philosophy and science to identify some key concepts and arguments that inform contemporary debates about causation. In Chapter 3 we will explore ideas of necessity and sufficiency, and see how they are related to regularities, laws and causes. In Chapter 4 we consider probabilistic causality, investigating the suggestion that causes raise the probability of their effects. Chapter 5 introduces counterfactual theories of causation, while Chapter 6 introduces the idea of a causal model, and considers a different kind of counterfactual theory built on the notion of experimental intervention. Chapter 7 concludes our tour of the standard theories of causation with a look at causal processes and mechanisms.

In Chapter 8 we take up the challenges of integrating agency and purpose into an account of the world's causal structure. We'll touch on how questions of causation impact our understanding of the relation between the mental and the physical, as well as the possibility of free will. Chapters 9 and 10 shift the focus from metaphysical and semantic questions to methodological questions about how we can find and use causes. Chapter 9 explores the role of observation and experiment in causal inference. It also discusses statistical and computational methods for finding causes in data, as well as techniques for integrating different kinds of evidence to establish causal hypotheses. Chapter 10 explores how we use causal knowledge to explain things, as well as to change them. We'll consider whether or not all explanation is causal, and we'll also identify principles that can be used to pick out the causes that are best able to help us explain, predict, and control phenomena.

While the chapters are most easily read in order, I have tried to make the text as modular as possible, so you can skip around—especially if material in the earlier chapters is already familiar to you. Theories of causation and causal inference often employ technical terminology, and there are occasionally unavoidable bits of logic and math. I've done my best to avoid technical notions when possible, and when I cannot avoid them, I try to stop and provide clear definitions and examples. There is also a glossary to aid you.

SUGGESTIONS FOR FURTHER READING

The online *Stanford Encyclopedia of Philosophy* has an extensive set of helpful entries on topics related to causation. The *Internet Encyclopedia of Philosophy* has less extensive coverage, but what it has is excellent and accessible. *The Oxford Handbook of Causation* (Beebee, Hitchcock, and Menzies 2009) is a comprehensive source, with entries by many well-known experts in the field. Illari and Russo's (2013) *Causality: Philosophical Theory Meets Scientific Practice* is, like this book, motivated by the desire to bring together philosophical and scientific approaches to causation, but contains somewhat more advanced and systematic discussions of the contemporary philosophical literature.

2

THINKING ABOUT CAUSATION: A SELECTIVE HISTORY

Causation has been a subject of scientific and philosophical reflection since antiquity. A serious overview of this history would require a book in itself, so in this chapter I must confine myself to a select few thinkers whose ideas have most directly influenced the contemporary accounts discussed in subsequent chapters.

ARISTOTLE

While reflection on causes can be found in the work of the earliest Greek philosophers, the first systematic account is due to Aristotle, writing in the first half of the 4th century BCE. Aristotle believed that to understand anything in nature, we must know its causes, and his theory of causes permeated his entire natural and practical philosophy.

Aristotle held that everything that exists has four kinds of causes, which have come to be called material, formal, efficient and final causes (*Physics* II 3; *Metaphysics* V 2). The distinction between these kinds of causes is best illustrated by means of an example. What, for instance, should we say caused the Parthenon? First, we might point to the material it is made of—the many blocks of marble. This is its material cause. We can also point to the Parthenon's plan—its arrangement of flooring, pillars, pediments, roofing, and so on. It is this arrangement that makes the Parthenon the thing it is and distinguishes it from the heap of

DOI: 10.4324/9781003200956-2

material from which it is made. This is its formal cause. Third, we might point to the masons and laborers who built the structure. they are the Parthenon's efficient cause. Finally, we might look to the purpose of or reason for the structure—in the case of the Parthenon this might be to honor Athena or to demonstrate the power and prestige of Athens. These reasons or purposes are its final cause. To completely understand why the Parthenon exists, Aristotle thought we must know all of these things.

Although it is plausible to think that a complete understanding of why something like the Parthenon exists demands attention to each of these factors, it is less clear to the modern ear that we should count all these factors as *causes*. To understand why the Parthenon exists as it is, we should know what it is made of and understand its plan, but these things are not events or happenings, and they are not producers of change.

It may therefore be better to think of Aristotle's four causes as four reasons or four ways to answer why-questions; indeed, the Greek word Aristotle is using for cause (*aition*) can also be translated as "reason." If we adopt this reading, Aristotle's main concern is **explanation**, and he suggests that complete explanations will require an account of the reasons why something is, along each of these four dimensions. This interpretation of Aristotle is reasonable so long as we are clear that, for Aristotle, these explanations are true features of the world—things that we can discover and things that we can be wrong about. If I tell my son that the present is under the tree because Santa brought it, then I would not really have offered an explanation, since I, and not Santa, was the efficient cause.

Contemporary philosophers generally hold that Aristotle's efficient cause corresponds most closely to our contemporary notion of cause. The correspondence is far from exact, and in fact, as we shall discover, it is far from clear that there is a single "modern" sense of cause. Still, efficient causes, which Aristotle described as "the primary source of the change or coming to rest" (*Physics* II 3 194b30), do indeed seem to be the kind of causes that fit best with both folk and scientific uses of the term "cause."

A factor that will loom large in later debates about causality is the status of final causes. Aristotle introduces final causes by means of examples involving human-made artifacts; final causes in these cases are just the purposes for which the artifacts were designed and built. If you are explaining why a car exists, you can point to its efficient cause—the activities of the machines and workers that built it, but you can also point to its purpose—to its capacity to get you around town. But Aristotle attributed final causes not just to artifacts, but to many things one finds in nature; not just cars, but coyotes and clouds, have final causes. For Aristotle, this was because everything had a nature, and with this nature comes tendencies or ends towards which change is directed.

DESCARTES

René Descartes (1596–1650) was one of the most influential mathematicians, scientists and philosophers of his generation. He is sometimes called the last medieval philosopher, and the first modern. Descartes had been educated within French universities dominated by the scholastic tradition. This tradition, which developed in the Middle Ages and received perhaps its fullest expression in the work of Thomas Aquinas, sought to provide an account of nature and of human beings that integrated ancient philosophical sources, especially Aristotle, with the philosophical and theological teachings of the Roman Catholic Church. In this Christianized Aristotelianism, Aristotle's account of the four causes was widely accepted, but the Christian element provided a somewhat revised conception of final causes. Aristotle was not a theist in anything like the Christian sense, but in the Thomistic version, it was easy to find a source of final causes in the will of the benevolent and omnipotent God.

Descartes, following in the footsteps of Galileo, was eager to reject the scholastic approach to science and to replace it with one that trusted to human observation and reasoning over ancient and ecclesiastical authority. And like Galileo, he was eager to overturn Aristotelian orthodoxies about matter and motion.

In its place Descartes advocated an account of the physical world that came to be called the mechanical (or corpuscular) philosophy. The heart of this mechanical philosophy, which was embraced in some form by many 17th-century philosophers/scientists including Robert Boyle, Robert Hooke, and Pierre Gassendi, was the idea that the observable properties of physical things—like their color, temperature, or smell—were actually the consequences of the shape, arrangement and motion of microscopic bits of matter. For instance, the heat of a macroscopic body was understood to depend upon the amount of microscopic motion within its parts or corpuscles.

While the particular mechanical hypotheses made by Descartes and others have generally not been born out, subsequent developments in the history of science have shown just how fruitful the mechanistic approach has proved to be. To name a couple of examples, contemporary thermodynamic theories do indeed equate heat with motion, and contemporary theories in chemistry and molecular biology explain phenomena like enzyme action or gene expression via the interlocking of molecules of different shapes or the unfolding or refolding of reticulated structures like DNA and proteins.

Cartesian mechanical philosophy brings with it a view about what can cause what. For Descartes, causation required physical connection, ultimately reducing to direct contact between corpuscles. Think of one billiard ball striking another, or of two interlocking gears where the teeth of one pull the second gear around in circular motion. These causal interactions look like efficient causes in Aristotle's sense. They are the activities which are producers of motion and rest.

Descartes did not deny that there were final causes, but thought they were beyond human ken:

> I know that my own nature is very weak and limited, whereas the nature of God is immense, incomprehensible and infinite [and]... that he is capable of countless things whose causes are beyond my knowledge. And for this reason alone I consider the customary search for final causes to be totally useless in physics; there is considerable rashness in thinking myself capable of investigating the purposes of God."
>
> *(Descartes 1984, 39)*

The proper kinds of causes to be studied by physics (and by extension all the sciences built upon them) involved contact between bits of matter in motion.

One of the most puzzling legacies of Descartes' mechanical philosophy was its implications for our understanding of mental causation—which is the problem of understanding how mental events and states, like our perceiving something, or being angry, or reaching a decision, can cause and be caused by happenings in the physical world.

Like many natural philosophers of his generation, Descartes sought an account of the mind and world that would allow him to accept the new science while holding on to his Christian faith. This project is nowhere clearer than in his *Meditations on First Philosophy*, the subtitle of which is "in which is demonstrated the existence of God and the immortality of the soul." Descartes argued that mind and body are distinct substances with distinct essential properties. A mind is a thing that thinks, while a body (including both the human body and physical bodies generally) is a thing that is extended—a thing that occupies space, has shape, and can be in motion or at rest. Since mind and body have different essential properties, a mind cannot be identical to a body, or to a part of a body like a brain.

This view, known as substance dualism, has the advantage that it allows the mind/soul to survive the death and decay of the body. But, on the other hand, if the mind is not part of the physical world, it is a mystery how goings on within the mind could cause or be caused by goings on in the body and the world surrounding it. The problem is particularly acute for Descartes, given his view that causation required action by contact. Descartes, who had engaged in extensive studies of human and animal anatomy and physiology, viewed bodies, both human and animal, as machines, though he thought only human bodies were directed by minds or souls.

Descartes argued that minds and bodies must causally interact; experiences in our mind (sights, sounds, pains, etc.) were caused by physical events in and beyond our bodies, while decisions of our mind caused voluntary motions of our bodies. The puzzling question was where and how this interaction between mind and body occurred. Descartes supposed that the location of this interaction was the pineal gland, a small gland located near the center of the human brain.

Few have found Descartes' account of mind-body interaction satisfying. How, if action requires contact, can something that does not occupy space, and has no mass and no location, touch or push something in the physical world? Given this challenge, philosophers have been pressed to offer alternative solutions, chiefly by arguing either that, contrary to appearances, minds and bodies do not interact, or by giving up on substance dualism and accepting that the mind is a physical thing, and hence something capable of acting in the physical world.

NEWTON

Sir Isaac Newton (1643–1727) was born shortly before Descartes' death, and like Descartes, was a mathematician, physicist, and philosopher. In his magnum opus, *Principia Mathematica*, Newton developed what has been come to be called Newtonian mechanics—a framework for describing the motions of physical bodies and the forces that act upon them. To describe these motions mathematically, Newton invented new mathematical techniques that we now call the calculus. (Leibniz, along with

Newton, is credited with developing the calculus; and the calculus that is taught today uses variations on Leibniz's formalism.)

Using these tools, Newton offered a theory of gravitation that solved the millennia-old problem of planetary motion, showing how Kepler's laws could be derived and corrected by supposing that all objects attract each other gravitationally proportionally to the product of their masses divided by the square of the distance between them. Newton's theory also provided a unified account of celestial and terrestrial motion. The very same laws used to describe and predict the motion of the planets, the sun, and the moon, could be used to derive principles like the law of falling bodies, to predict the path of projectiles like cannon balls, and to explain the patterns of the tides. Not without reason is *Principia Mathematica* often considered the greatest work in the history of science.

Newton's achievement had several important consequences for philosophical accounts of how causes acted in the natural world. For one thing, Newton's account describes causal dependencies using mathematically precise and general statements of laws. Take the law of universal gravitation, which in contemporary notation is written as $F_g = Gm_1m_2 / r^2$. Here G is a constant (the constant of universal gravitation), m_1 and m_2 are the masses of two objects, and r is the distance between those two objects. This law specifies a mathematically precise measure of the instantaneous force holding between every object in the universe. Forces acting on bodies will accelerate those bodies towards each other according to Newton's second law ($F = ma$). Using the law of universal gravitation together with his laws of motion, Newton was able to derive further laws, like Kepler's laws and the law of falling bodies.

Forces can also be combined. While the gravitational attraction between the earth and the sun can account for the earth's basic elliptical motion around the sun, a more precise trajectory can be calculated by adding gravitational forces coming from smaller or more distant bodies, like the force between the earth and the moon, or between the earth, sun, and neighboring planets. And not all forces need to be gravitational. For instance, idealized models of gravitational systems, like falling bodies or bodies rolling down inclined planes, can be made more precise by incorporating forces of friction and air resistance. This capacity to add together causal influences in mathematically precise ways gave Newtonian mechanics an exceptionally powerful set of tools to represent the world's causal structure.

Newton's scientific successes seem to vindicate the principles of mechanical philosophy—offering a glimpse of a universe operating with clock-like precision. But they also pose a major challenge for mechanical philosophy in the Cartesian style. Descartes' account of causation imagined that all causal interactions require contact. That meant that for one object to act upon another, they must be adjacent. But according to Newton's theory of gravitation, every massive body in the universe acts upon every other to some degree, no matter how far away they are. This is the problem of action at a distance.

Descartes considered action at a distance to be unintelligible; it required believing in occult forces. He reasoned therefore that there must be some mechanical cause of gravity. To account for the motion of the planets he supposed that the heavens were filled with a fine celestial matter which carried planetary bodies in orbits in the same way that eddies in a stream might carry a stick in a circular pattern. To account for the tendencies of bodies to fall towards earth, he imagined them to be pressed downward by the celestial matter in the way that an object might be carried by whirling water towards a drain at the center of a sink.

Newton was not impressed by Descartes' argument. Descartes had not provided a mathematical description of the vortices that could propel planets in their orbits and Newton saw insuperable difficulties in offering such an account; moreover, Descartes had no way to detect this celestial matter. Contrasting his own approach with Descartes', Newton wrote, after describing his inverse square law:

> I have not as yet been able to deduce from phenomena the reason for these properties of gravity, and I do not feign hypotheses. For whatever is not deduced from the phenomena must be called a hypothesis; and hypotheses, whether metaphysical or physical, or based on occult qualities, or mechanical, have no place in experimental philosophy. ... [I]t is enough that gravity really exists and acts according to the laws that we have set forth and is sufficient to explain all the motions of the heavenly bodies and of our sea.
>
> *(Newton 1999, 943)*

While Descartes was not the last to seek a causal explanation of gravity, the empirical success of Newton's theory led Newton's view to prevail—at least until Newton's theory was succeeded in the twentieth century by Einstein's theory of general relativity—a theory that ironically does away with action at a distance.

The importance of Newton's legacy is twofold. First it connects causation to laws and regularities; Second, it liberates thinking about causal processes from the requirement of connection by pushes and pulls. Hereafter we are open to considering causal explanations that appeal to diverse forces. There may be more fundamental forces besides gravity—for instance, electricity and magnetism. And equally important, we may generalize our notion of forces or activities to other kinds of causes, from chemical bonds to psychological drives.

HUME

The influence of David Hume (1711–76) on philosophical and scientific thinking about causation is difficult to overstate. His analysis of the epistemology and metaphysics of causation was transformative, and has set the terms of all subsequent discussion. Hume was an empiricist, much impressed by the experimental philosophy of Newton and the other physicists of his day. Broadly speaking, Hume's aspiration was to apply this approach from the physical sciences to the study of human nature. Hume recognized that the study of the human mind could not admit of the mathematical precision of Newtonian physics, and Hume was not what we would call an experimentalist; but he did insist that all our knowledge of the world came from our experience of that world, and that, to properly understand concepts like causation, we must attend carefully to how our experience gives rise to those concepts.

Hume argued that human knowledge can be divided into two kinds—relations of ideas and matters of fact. Knowledge about relations of ideas is the kind of knowledge we get from logic and mathematics, and its key feature is that this knowledge does not depend upon our experience of the world. To come to know the Pythagorean theorem, for instance, we need not (and indeed cannot) rely upon measuring the lengths of sides of triangular objects in the world; we must instead prove that it holds by examining the relevant definitions. But, on Hume's view, all knowledge of matters of fact, of goings on in the world, must come from our experience of the world—what we see, hear, feel, and so on—together with our knowledge of cause and effect. How for instance do I know my brother is in Washington? I know this because he sent me an e-mail that told me so. I suppose that the e-mail I received was caused by actions that he took, and that his action of reporting his trip to Washington was caused by his belief that he was in Washington, and that this belief was caused by actions which he took to get him to Washington, and so on .... Without making inferences about cause and effect I can make no judgments about any fact not immediately before me—whether it be where David Hume is buried or what my next sip of coffee will taste like. Given the centrality of cause-and-effect reasoning to all of our knowledge, Hume felt it necessary to discern the source of our ideas of cause and effect, and the kind of justification we can have for inferences based upon cause and effect.

Hume understood all human experience to consist of two categories of perceptions—ideas and impressions. What he called impressions we might most naturally refer to as sensations, like the warmth of a candle or the redness of an apple, though Hume also took them to include inner sensations, like a feeling of fear or hunger. While there is much more to what we experience than these impressions, Hume thought that everything else (the ideas) originally derive from them. If you are sipping a cup of coffee right now, you have immediate impressions of its taste, its smell, its color and so on. But even if you are not right now sipping coffee, you still have an idea of coffee that you can conjure up based upon your past impressions. And you can compound your experience in many different ways. For instance, you can conjure up an idea of a purple goat, even though I doubt you have ever seen a purple goat. You have seen goats, and you have seen purple things, and that is enough. Hume thought that even the most abstract ideas derive ultimately from the impressions of sensory experience. Take for instance the idea of God as an omnipotent being. Omnipotence is not something we have experienced, but we do know what it is NOT to be omnipotent—to be unable to lift something, to be unable to change the world in some way—and that is enough to conceive the idea of God as a being without such limitations.

Thoughts for Hume were associations of ideas—for instance a thought of coffee will include a bundle of ideas of taste, smell, texture and so on. And Hume maintained that there were only three kinds of relationships by which those ideas could be associated—resemblance, contiguity in time and place, and cause and effect. My idea of an apple is a bundle of simpler ideas (color, shape, smell, etc.) contiguous with each other in time and place. I take two objects both to be apples because my bundles of ideas of them resemble each other with respect to their color, shape, smell, etc. And if I take a bite of the apple and then taste the sweetness and crunch, I take my biting into the apple to be the cause of that taste.

Hume observed that when we take two things to be related as cause and effect, we believe there to be a kind of *necessary connection* between them—some force or power by which the cause brings about or produces the effect. But from whence do we derive this idea? Hume's crucial claim is that this idea cannot come from any single impression, but only can arise as a result of observing a constant conjunction between contiguous events. We believe for instance that breathing underwater will suffocate us, but there is nothing in the sensible qualities of water—its color, its temperature, its wetness—that implies this. We only learn this connection as the result of experience—of the putting of our head underwater as being immediately and repeatedly followed by our experiencing an inability to breathe.

Hume thought this same principle applied even to the most basic of causal interactions. If one billiard ball strikes another straight on, the first one slows and the second one begins to move. We expect this, and assume the motion of the one produces the motion of the second; but, Hume argues, there is nothing in our impressions of the two balls that entails that this is what will happen. We can conceive that something else might happen—that the balls might head in a different direction, that they might fuse, or even that they might disappear. Hume concludes from this thought experiment that there is no impression of a "secret connection" between the two balls and that our conviction that the ball's striking *causes* the subsequent motion arises only from observing repeated instances of such strikings being followed by such motions.

On the basis of this analysis, Hume suggested a novel definition of cause. It is worth quoting his words in full:

> Similar objects are always conjoined with similar. Of this we have experience. Suitably to this experience, therefore, we may define a cause to be [1] *an object, followed by another, and where all the objects similar to the first are followed by objects similar to the second.* Or, in other words [2] *where, if the first object had not been, the second never had existed.* The appearance of a cause conveys the mind, by a customary transition, to the idea of the effect. Of this also we have experience. We may therefore, suitably to this experience form another definition of cause, and call it, [3] *an object followed by another, and whose appearance always conveys the thought of another*
>
> *(Hume 1975, 76–77)*

In these definitions Hume uses the word "object," but it is clear from the context that he principally is thinking of events. So a cause would be an event like a cue ball striking a second ball, and the effect would be the event of the cue ball slowing and the second ball accelerating.

Hume claims to have offered two alternate definitions of cause, but there really are three, indicated by the numbers I've added to the text. Definition [1] is the wellspring of regularity theories of causation, which for much of the three hundred fifty years since Hume wrote these words has been the predominant approach to causality amongst scientists and scientifically minded philosophers. Definition [2] is, despite Hume's "in other words," not a paraphrase of definition [1], but expresses the kernel of a counterfactual theory of causation, which is widely seen as an alternative to regularity theories.

Definition [3] is a subjective definition of cause, for it makes the fact that one thing causes another depend, not upon anything in the objects themselves, but upon the psychological convictions of someone observing the cause and the effect. Definition [1] is closely allied to definition [3] because Hume believes that it is only through the repeated appearance of similar patterns of contiguous objects/events that we come, by force of habit, to anticipate the effect when we see the cause, and to *feel* the sense of necessity. Hume emphasizes that our tendency to anticipate effects on the basis of causes is not founded in reason. We are conditioned to expect the effect to follow the cause in much the same way that Pavlov's dogs were conditioned to expect food after the sound of the bell. Like those dogs, we don't understand what is going on behind the scenes, and so we cannot rationally form expectations of a supposed effect based upon the appearance of a cause.

Hume's views on causality are much subtler than I am able to convey in this brief summary, and scholars remain divided on how best to interpret some of Hume's views. One disagreement is worth mentioning here. On the standard interpretation, Hume is seen to be denying that there really is anything like necessary connection in the world; all there is in the world are regularities—patterns of happenings, one thing after the other. Orthodox regularity theories in the 20th century followed this approach. An alternate view has Hume drawing only a more modest epistemological conclusion. On this view, Hume is not denying the existence or intelligibility of a "secret connection." He is merely saying that the ultimate character of causal powers is unknowable to us, so that causes can only be inferred on the basis of regularities.

KANT

Immanuel Kant (1724-1804) credited Hume with rousing him from his "dogmatic slumber." Hume's skepticism about causation and other metaphysical concepts raised the specter that the metaphysics of Kant's day was, as Hume put it in his *Enquiry*, all "sophistry and illusion." Kant was sympathetic to much of Hume's critique, but he thought that his skepticism was too far reaching. Among other things, Hume's critique of laws of nature and causation seemed to undermine the rational basis for scientific inquiry. Kant's critical philosophy aimed to delimit the proper bounds of metaphysics, setting out what questions were and were not within the scope of human understanding, as well as to provide an epistemological foundation for Newtonian science.

The starting point of Hume's skepticism was his argument that all concepts were derived from experience. Kant rejected that argument, contending that some principles were knowable *a priori* (independently of experience), among them the principles that every event has a cause and that causes necessitate their effects. Kant argued, contra Hume, that our concept of cause, as well as other basic concepts, like our concepts of space and time, are not derived from experience, but instead are preconditions of the very possibility of experience. Because we in fact do have experience, and because we could not possibly have any experience at all except through these concepts, these concepts *must* be applicable to that experience. All of our experience is, of necessity, experience of happenings in space and time, and experience of things causing other things.

While Kant believes this kind of argument (called a transcendental argument) secures the legitimacy of these concepts, it is crucial to his larger project to argue that these concepts may only be rightly applied within the bounds of experience. Kant draws a sharp distinction between the realm of *phenomena*—things as they appear to us in our experience—and the world of *noumena*—or *things in themselves*. Everything within ordinary perceptual experience, as well as everything that is a conceivable subject of scientific investigation, falls within the realm of phenomena. But these phenomena are not things in themselves, but are instead things as they are presented to us in our intuition and understanding. Concepts like space, time, and causation are in a sense imposed upon the world to constitute experience, and we are therefore able to depend upon their universal validity within the realm of experience.

At the same time, there are many questions treated in metaphysics that are beyond the realm of experience—for instance, the existence of God, freedom of the will, and the immortality of the soul. These matters lay outside the scope of our perceptual and scientific concepts. It is perfectly well, thought Kant, to recognize the necessity of a causal relation between objects in the world, like the gravitational pull of the earth upon the moon, because these causal relations obtain in the phenomenal world. But to ask whether and how God caused the world to be is to use the concept of cause in a place it has no business going. Kant's argument was a brilliant maneuver, because it seemed to save at a stroke, science, religion, and morality. Human freedom and moral responsibility are of the noumenal realm, and so we cannot and need not reconcile it with the causal structure of the phenomenal world.

All of this is just a very rough sketch. Kant's *Critique of Pure Reason (1965)*, in which these arguments appear, is among the most difficult works in the history of philosophy, and open to a variety of interpretations. What is most important for our purposes is this: Kant's critical philosophy sought both to legitimate causal reasoning and to delimit its proper realm. And in so doing he sought to pull human will and agency from the world's causal fabric. Whether this move is ultimately tenable is a matter of continuing debate.

MILL

John Stuart Mill (1806–73) is most widely known as a moral philosopher and political theorist, but he also made important contributions to logic and the philosophy of science. His massive *System of Logic* includes an account of the nature of causes and laws of nature, along with a set of principles for causal inference that have come to be called "Mill's methods."

Mill was an empiricist whose thoughts on causation owed much to Hume. But unlike Hume, he did not think that empirical principles inevitably lead to skepticism. Mill took it as axiomatic that there are uniformities in the course of nature, and that it is the fact of these uniformities that makes scientific inference possible. Mill's view of scientific method was fundamentally inductive; he believed scientists gain knowledge by observing particular instances and then articulating laws describing patterns in the instances.

Like Hume, Mill adopted a regularity theory of causation in which causes are antecedent conditions that are sufficient for their effects. He defines "the cause of a phenomenon to be the antecedent, or the concurrence of antecedents, on which it is invariably and unconditionally consequent" (Mill 1874, 245). This definition ties together the notions of cause and law. What Mill calls the "law of causation" is just the principle that "every event depends on some law" (Mill 1874, 236). We can put the law of causation simply as "same cause, same effect."

Mill speaks of the "concurrence of antecedents" because seldom or ever is one antecedent condition sufficient for an effect. Matches don't always light when struck, acorns do not always become oaks, and sometimes the bread does not rise. To get a true cause we must include not just a single condition, like the striking of the match, but all the other conditions that must be in place. The surface must be rough enough to induce a spark; the match needs to be dry, the wind gentle enough, and so on. Only given all of these conditions (the total cause) does the effect invariably follow.

Mill described four methods for finding causal laws. The first is the method of agreement. If we have some phenomena p that has occurred repeatedly, the method of agreement counsels us to consider each of the cases of p, and find the common factor which they all share. A good example of this method is familiar from the procedures for tracing food-borne illnesses like *E. Coli* infections. Suppose we have an outbreak of these infections at approximately the same time across seven different states in the US. The victims of the illness each of course have eaten many different foods in the days prior to their infection. To figure out which of these foods are the cause, we seek to identify a particular food from a particular food source. All of them, we discover, ate romaine lettuce that came from a single farm in California. We infer from this that the source of the infection is that food.

Mill's second method is the method of difference, and here, instead of looking for the common factor in a set of cases of a phenomenon p, we compare circumstances where p occurs with otherwise similar circumstances where p does not occur, and look for some factor that is present in all of the p cases and none of the not-p cases. This factor we take to be the cause. As an example of this, consider again food poisoning. Suppose that twenty out of one hundred diners at a restaurant exhibited symptoms of food poisoning. We analyze their menu choices and discover that all twenty who had gotten sick had ordered the seafood omelet, and that none of the other eighty had done so. We conclude that the seafood omelet was the cause of the food poisoning.

Mill discussed two other methods—the method of residues and the method of concomitant variation. The method of residues is really derivative upon the method of difference, and need not trouble us here, but the method of concomitant variation deserves a mention. Briefly put, that method counsels us that whenever some quantity can be varied and another quantity systematically covaries with it, we may infer that one is the effect of the other. Suppose for instance that we turn the dial on a speaker, and the volume increases and decreases systematically as we move it clockwise and counterclockwise. We are justified in inferring that the turning causes the volume changes.

What is important about this method is that it moves us from qualitative to quantitative causal relationships. The methods of agreement and difference focus on causes and effects that are events, and events either occur, or they don't. But many of the causal relationships of interest to us are quantitative. We want to know what made the tree grow taller, or what caused the patient's pulse to speed up, or what caused the image on the screen to get cloudier. Quantitative causal relationships are essential in science, though they often have received less attention than they should from philosophers.

Mill calls these methods "experimental," but in his view they did not really require experiments, and our examples illustrating the methods of agreement and difference bear this out. In the examples I discussed to explain the methods of agreement and difference, no experiments were performed. All that we imagined doing was observing the varying effects that we saw associated with varying combinations of possible causes.

What distinguishes experiment from observation is that in experiments we do not simply observe, but instead intervene in the world and then observe. But we can live without intervention so long as we have sufficient variation in conditions that we can see agreement and difference among causes and effects. These variations create what are called "natural experiments." Take for instance the food poisoning. People's menu selections have created two different conditions—the omelet-eating condition and the no-omelet-eating condition. This is very much like an experiment, except that the experimenter didn't choose who ate the omelets. Generally speaking, these natural experiments provide strong evidence of causal relationships, though there are caveats we shall address in Chapter 9.

RUSSELL

Bertrand Russell (1872–1970) was one of the most influential philosophers of the twentieth century. He rose to prominence because of his work in logic and the foundations of mathematics, and he is generally seen as one of the founders of analytic philosophy. Russell was instrumental in turning much of philosophy away from earlier styles of metaphysics and towards conceptual analysis. Russell's principal contribution to the philosophy of causality comes from a short essay "On the Notion of Cause," published in 1913. Russell believed that the concept of cause was hopelessly confused, and that both science and philosophy would be better off without it. In the most famous passage of the essay, he quips:

> The law of causality, I believe, like much that passes muster among philosophers, is a relic of a bygone age, surviving, like the monarchy, only because it is erroneously supposed to do no harm.
>
> *(Russell, 1913, 13)*

The law of causality that Russell rejects is roughly Mill's principle of same cause, same effect. One important problem Russell raises is that there is a great deal of vagueness in what counts as the "same cause" or "same effect." If we interpret "same cause" in a coarse-grained way, we will often find that the same cause is not followed by the same effect. If, for instance, we consider striking a match to be a cause of it lighting, we know that strikings are not always followed by lightings. If, on the other hand, we describe exactly the whole set of conditions required to guarantee a match's lighting, it will include much more than the mere fact of the match's striking. Indeed, it may have to include so much information about the particular features of a match and its environment, that the same conditions will never occur more than once.

In comparing causality to the monarchy and calling it a relic, Russell is suggesting that as our scientific understanding of the world increases, our need to speak of causes will diminish. He points out that the laws that one finds in sciences like physics aren't typically causal generalizations like "Whenever *C* occurs *E* occurs." Russell explains:

> The law of gravitation will illustrate what occurs in any advanced science. In the motions of mutually gravitating bodies, there is nothing that can be called a cause, and nothing that can be called an effect; there is merely a formula. Certain differential equations can be found, which hold at every instant for every particle of the system.
>
> *(1913, 13–14)*

While Russell's example is drawn from Newton, he believed that this was the direction towards which science in general was increasingly headed. Writing at the time of the birth of special relativity and quantum mechanics, physics seemed to be developing ever more precise and general mathematical laws, expressed as equations. These equations are difficult to interpret causally because many of them are temporally symmetric. While causal generalizations have a direction (causes precede effects), these equations do not. This is why you can equally well use the equations of Newtonian mechanics to predict the next solar eclipse and to "retrodict" the last one.

Russell's critique of causality proved influential. The logical empiricist movement in the philosophy of science, which was inspired both by the empiricism of Hume and Mill and by new developments in logic from Russell and others, developed an account of scientific theories in which causes largely disappear. Physics was taken as the paradigm science, and it was expected that as other sciences like chemistry, biology and psychology advanced, they too would replace vague causal talk with precise quantitative descriptions of regularities within their domains. Moreover, since it was generally agreed that the material basis of all natural phenomena is physical, it was expected that theories in these sciences could ultimately be reduced to physical theory. If one could provide definitions of chemical entities and processes in the language of physics, biological entities and processes in the language of chemistry, and so on, one would get a unified picture of science and nature, in which causes fall away, and in which all regularities in the world are ultimately physical regularities.

But this attempt to banish causes from physics, and by extension, other sciences, has not ultimately proved successful. While many physical laws are time-symmetric, not all are. One critical exception is the second law of thermodynamics, which implies that physical processes, in the absence of work, will evolve towards maximum entropy. This is the principle which tells us that heat moves from warmer to cooler bodies and it explains (in a roundabout way) why living things need energy to survive, and why you can't simply reverse the physical process that scrambled the egg in order to unscramble it. It remains a major theoretical question in physics how to understand how the temporally asymmetric principles of thermodynamics are connected to symmetric dynamical laws found in physics. At the moment it is unresolved what grounds the asymmetries we see in thermodynamics, in cause and effect, and in time itself.

Leaving these theoretical puzzles aside, examination of the practice of physics suggests that causal reasoning and manipulation remains essential. Everything we believe about the fundamental structure of the physical world is grounded in some sense on knowledge gained by experiment. But experiment, in physics and elsewhere, is at root a causal procedure. It involves intervening in the world and detecting the intervention's effects. As Nancy Cartwright (1979) has argued, asymmetric laws relating cause and effect seem essential for getting around in the world.

If causation remains alive in physics, it remains even more so in the other sciences. This is in large part because phenomena described by the life and social sciences do not appear to be governed by the sort of simple temporally symmetric functional relationships that Russell pointed to in physics. Recent work in philosophy of science suggests that much of what scientists find in these domains are not laws, but mechanisms—and mechanisms are, as we will discuss in Chapter 7, essentially causal.

SUGGESTIONS FOR FURTHER READING

A reader interested in the history of philosophical discussions of causation could do no better than to begin with the short passages from Aristotle's *Physics II*, Chapter 3 and Hume's *Enquiry Concerning Human Understanding*. Russell (1913) is also valuable and accessible. Mackie's (1980) *The Cement of the Universe* is a classic in the contemporary causation literature, and contains substantial analyses of Hume, Mill, and Kant. Strawson's (2014) *The Secret Connexion* provides an extensive if controversial analysis of Hume's views.

3

CAUSES, REGULARITIES, AND LAWS

In this chapter we will explore a family of approaches to causation that descend from Hume and Mill, and which dominated much of twentieth century thinking about causation. We begin by looking at how ideas of necessity and sufficiency can be analyzed in terms of conditional statements, and how these conditionals can be used to interpret causal claims as claims about difference-making. We'll then introduce a simple regularity theory of causation inspired by Hume's first definition of a cause. We'll identify a number of difficulties with this theory, and investigate how more sophisticated regularity theories seek to evade them. We'll conclude by discussing the relationship between causes and laws of nature, and briefly discussing two theories of what laws of nature are.

THE CONDITIONAL ANALYSIS OF NECESSITY AND SUFFICIENCY

Our ordinary understanding of the words "necessary" and "sufficient" can be expressed using conditional (if…then) statements. To say that an event x is **sufficient** for event y is to say that if x occurs, then y must occur. To pass my logic class, for example, it is sufficient to achieve an average on assignments of 60%. If you get to 60%, then you will pass the class. Necessity is the other way around. To say that an event x is **necessary** for an event y is to say that if y occurs then x must have occurred,

DOI: 10.4324/9781003200956-3

or, equivalently, if *x* doesn't occur then *y* won't occur. To pass my logic class, for example, it is necessary that you take the final. If you don't take the final, then you won't pass.

To spell out exactly what is entailed by these claims we must understand the conditions under which if...then statements are true. The standard analysis from propositional logic interprets such statements as **material conditionals**. Here's how these conditionals work. The statement:

If John gets an average of 60% or better, then John passes the logic class.

is a conditional. It has within it two components. The first, in the "if" phrase, is the antecedent:

A: John gets an average of 60% or better.

The second, in the "then" part, is the consequent:

L: John passes the logic class

To render the original conditional as a statement in propositional logic, we assign each of the components a letter as a name, in this case "A" for the antecedent and "L" for the consequent. We then symbolize the original conditional by joining the antecedent and consequent with the horseshoe symbol representing the material conditional:

$A \supset L$

Here is how in general we translate necessary and sufficient conditions into material conditionals:

	Conditional	*Symbolization*
p is sufficient for q:	if *p* then *q*	$p \supset q$
p is necessary for *q*:	If *q* then *p*	$q \supset p$

Notice that the use of material conditionals has shifted our analysis of necessity and sufficiency from events to statements. Events are happenings in the world, while statements are sentences that are true or false. John's passing the logic courses is an event, while "John passed the logic course" is a statement asserting that this event occurred.

Given this analysis, we can now state the conditions under which statements that *p* is necessary or sufficient for *q* are true. The horseshoe is a truth-functional connective, because the truth value of a statement formed with it is a function of the truth values of its component statements. The truth conditions for the material conditional can be represented in a truth table like this:

p	q	$p \supset q$
T	T	T
T	F	F
F	T	T
F	F	T

What this table indicates is that a conditional statement is true unless it has a true antecedent and a false consequent. Let's apply this rule to the claim that getting 60% on your test is sufficient to pass the course. This is equivalent to the statement "if you get 60% on your test you pass the course." The claim will be false if you do get 60% or more and don't pass the course; otherwise, it will be true.

The problem with using the material conditional to analyze causal necessity and sufficiency is plain. Because the truth of material conditionals depends only upon the truth values of their antecedents and consequents, a material conditional can be true even if there is no actual connection between the two. Consider for instance these two conditional statements:

If the Red Sox are a baseball team, Cambridge is a university.
If porcupines are native to Antarctica, Darwin is Polish.

If we treat these statements as material conditionals, then both are true. The first statement is true because both its antecedent and consequent are true (line 1 of the truth table), and the second is true because both its antecedent and consequent are false (line 4 of the truth table). But evidently the fact that the Red Sox are a baseball team has nothing to do with Cambridge being a university, and Darwin's nationality depends in no way upon the native continent of porcupines.

Clearly then, the truth-functional analysis is not by itself adequate to capture the necessity that holds between causes and effects. If I tell students that getting 60% is sufficient to pass the class, I am implying that their passing the class is **connected** to their passing that threshold. That is what they must do to pass, and that is enough. I am making a causal claim.

Without some connection, we might have one thing follow another, but we won't have causation. Something more is required to distinguish causes from things that are merely contiguous or correlated. Humeans claim that something is regularity.

HUMEAN REGULARITY THEORY

Here again is Hume's first proposed definition of a cause:

> we may define a cause to be *an object, followed by another, and where all the objects similar to the first are followed by objects similar to the second.*
>
> *(Hume 1975, 76)*

Let's consider the elements of this definition one by one. First, causes are "objects." For the moment let's treat these objects as events—specific happenings at some place in space and time, like a billiard ball striking another or the new president being elected. Second, there is the requirement that the first event, the cause *c*, is followed by a second event, the effect *e*. "Following" is a matter of temporal succession; Effects follow their causes in time. It is not clear in this passage to what degree following requires spatial contiguity. Third, we have the requirement that all events similar to the first are followed by events similar to the second.

This third element, the regularity requirement, assumes the idea of similarity between events. Similarity though is a slippery notion. Take any two events at all, and you can probably find some similarities between them. Joe Biden's inauguration is similar in some respects to Donald Trump's inauguration, but also in some other respects to Napoleon's coronation or Joe Biden's wedding. Perhaps a better way to describe the regularity is in terms of generic event types: for instance, Joe Biden's and Barack Obama's inaugurations are both single instances of the US presidential inauguration event type. While what is necessary for two events to be of the same type is not completely clear, the notion of an event type can at least help pick out the respects of similarity that are relevant to the regularity.

Using lower case letters to represent singular events and capital letters to represent generic event types, we can rephrase Hume's definition as a theory of causation I'll call Humean Regularity Theory:

Humean Regularity Theory (HRT):
An event c causes event *e* if, and only if,

1. *c*, of type *C*, occurs
2. *e*, of type *E*, follows *c*
3. All *C*s are followed by *E*s

HRT stipulates that causes precede their effects. This assumption is clearly part of our commonsense understanding of cause, and I shall accept it throughout this book, but it is worth noting that it can be challenged. For one thing, if time travel is possible, it would seem that events in the future could cause events in the past. For another, some philosophers have suggested that causal order may be more fundamental than temporal order, and have tried to define temporal order using causal order. If such a move is to work, one would not be able to make temporal priority a precondition for something to be a cause of some effect.

The last clause represents HRT's answer to the nature of causal necessity. What distinguishes cause and effect from mere coincidence is that cause-effect pairs are instances of exceptionless conjunctions between events of their type. HRT interprets the "must" of causation as "always." While I've stated HRT as a theory of singular causation, it has a simple generic form: *Cs* cause *Es* if, and only if *C*s are always followed by *E*s. And notice that even for singular causal claims, it is the general claim that grounds the singular one. A particular *c* causes a particular *e* only in virtue of all *C*s being followed by *E*s.

HRT implies that causes are sufficient for their effects but not that they are necessary for their effects. This is plausible because a generic type of event will typically have more than one possible type of cause. Headaches for instance can be caused by loud noises, head trauma, fevers, and stress, among other things. Clearly it would be wrong to insist that any one of these potential causes is necessary for a headache.

While HRT is a literal reading of Hume's first definition, it almost certainly does not represent Hume's considered view. The difficulty, of which Hume appeared to be aware, is that many of the things we identify as causes are not by themselves sufficient for their effects. We should certainly call having sex a cause of pregnancy. Yet, while it is necessary for pregnancy (excepting methods like artificial insemination and in vitro fertilization), it certainly isn't sufficient, since most cases of sexual intercourse do not produce pregnancies.

Failures of sufficiency often lead regularity theorists to suggest that most factors cited as causes are in fact only partial causes, and that what is sufficient is not a partial cause but a total cause. Sex is only a partial cause of pregnancy, since it is but one of a number of conditions that must obtain for a woman to become pregnant. For sex to result in pregnancy, the woman and the man must be fertile, the woman must be at the right point in her menstrual cycle, there must not be interfering forms of birth control, and a whole variety of physiological conditions must obtain in order for a sperm to fertilize an egg, and for the resulting zygote to implant in the uterine wall.

Even apparently complete causes can sometimes be prevented from causing their effects. For instance, snake bites from coral snakes are generally fatal to humans, but death can be prevented by administering the right anti-venom in a timely manner. This example points to a general problem. For any set of causes, no matter how complete, since these causes precede their effects, there may be possible interventions that can alter or stop the effect. That's what the anti-venom does.

One way to accommodate regularity theories to these sorts of cases is to treat the causal generalizations not as strict generalizations but as *ceteris paribus* (all things being equal) generalizations. These kinds of generalizations are particularly prevalent in the social sciences. For instance, *ceteris paribus*, raising interest rates decreases stock and bond prices while increasing bond yields, and *ceteris paribus*, winner-take-all voting systems will create two-party governance, while proportional voting systems will create multi-party governance. Such generalizations imply that their antecedents tend to cause their consequents, but allow for exceptions in cases where the effect is blocked or counteracting causes occur. For instance, even if it is not strictly the case that all coral snake bites are fatal, *ceteris paribus*, they are.

Another way to address these challenges is to treat the generalizations relating cause to effect as probabilistic. Perhaps we should think of causal laws as characterizing ways in which causes increase the likelihood of their effects. Certainly, sex increases the likelihood of pregnancy and coral snake bites increase the likelihood of death. We will explore how such a theory could work in the next chapter.

Most of our worries so far have been with cases in which the regularities are less than perfect. But what if there are no regularities at all? Hume's account presumes that the kinds of events that are causes and effects are events of types that are continually repeated. We have frequent experience of objects falling to the ground, of billiard balls colliding, or of being unable to breath when we put our head under water, but there are many cases of causes where the cause and effect seem unique. To take an obvious example, we think that the big bang is the cause of the universe's expansion, but there is no general regularity (accessible in our universe) of big bangs being followed by universe expansions. And the kinds of things we take to be causes of events in human or natural history—an extinction or speciation event, the loss of a battle, or the production of a work of art—often do not, to all appearances, instantiate some general regularity. We will need something better than HRT to handle such cases.

INUS CONDITIONS

We have seen that there are reasons to think that causes are not generally either necessary for their effects, or sufficient for their effects. But an account of causation based upon conditions of necessity or sufficiency might be repaired by combining these relationships in a more nuanced form. J. L. Mackie (1965) developed a proposal along these lines, arguing that causes are **I**nsufficient but **N**ecessary parts of **U**nnecessary but **S**ufficient conditions, or INUS conditions for short.

Here's an example close to Mackie's original to explain the idea. Suppose there is a house fire, and the inspectors identify the cause as a short circuit. Why should this be called the cause? The short circuit, by itself isn't sufficient for the fire; other conditions were required, like the presence of flammable material at the place where the short occurs. Additionally, the short circuit wasn't necessary for the fire either, because the fire could have been caused by other means—for instance by someone dropping a cigarette while smoking in bed, or by a grease fire in the kitchen. But what we can say of the short circuit is this: it was a necessary part of a conjunction of conditions (including positive conditions like the presence of flammable material and negative conditions like the absence of a sprinkler system) that taken together were sufficient to cause the fire.

More generally, for any event that might occur, there will be a possibly large set of combinations of circumstances, each of which, if any of them were to occur, would be sufficient for the event to occur. We can express the relation between causes and effects in quasi-logical notation schematically this way:

> **INUS condition:** If [(A1 & A2 & A3 &...) or (B1 & B2 & B3 & ....) or (C1 & C2 & C3 & ...) or ...] then E

The various letters here can stand for different kinds of propositions—like the proposition that an event occurred (e.g., the short circuit), or that a condition was present (e.g., there were flammable materials) or that something was absent or omitted (e.g., that the sprinklers were turned off). E is the proposition that the effect *e* occurred. Logically, the INUS condition is a conditional statement, in which the antecedent is a disjunction of sufficient conditions for E, with each sufficient condition consisting of a conjunction of necessary conditions.

If Mackie's analysis is correct, when we assert that a particular event *c* caused another particular event *e*, what we mean is that:

1) *c* and *e* occurred
2) the proposition that *c* occurs is a conjunct within one of the sufficient conditions for *e* occurring,
3) All of the other conjuncts within that sufficient condition are also true.

I've framed this as an analysis of singular causal claims, but a similar analysis works well for generic causal claims about types of events. The analysis seems to comport well with commonsense conceptions of causation and seems to give the right answers to examples we've discussed so far. Snakebites are one way to die on a camping trip, but there are others; and the bite alone is not sufficient, because certain conditions, like the absence of anti-venom must also hold true. Similarly, sex causes pregnancy, not because it is the only way to cause pregnancy (there are others ways like IVF), and not because sex alone is sufficient for pregnancy (since many other conditions must be right), but because it is one part of a set of conditions that taken together are sufficient for pregnancy.

While Mackie's account provides a useful analysis of the relationship between necessity and sufficiency in causal claims, its exact meaning won't be clear until we explain how necessity and sufficiency should be understood. If we are treating INUS conditions as analyses of generic causal claims, it seems possible that we can understand necessity and sufficiency, as Hume did, as universality: whenever all the conjuncts of one of the sufficient conditions occur, the effect will occur. But this analysis won't work very well for singular causal claims. Given that a particular house burned down, it is trivially true that whenever the house burned down (which happened exactly once), there was a short circuit. Equally, if it turned out there were a grapefruit in the fridge when the fire started, it would be trivially true that whenever there was a grapefruit in the fridge, the house burned down. We need something stronger to capture the notion of necessity involved.

It is more plausible to understand the conditionals in singular INUS conditions as subjunctive conditionals. When we say the short circuit was necessary for the fire, we mean that if the short circuit hadn't occurred, the fire wouldn't have occurred. When we say that the set of conditions (the short circuit, the flammable materials, etc.) that correspond to the total cause are sufficient for the fire, we mean that if they were all to occur (as they did on this occasion), then the fire would occur. The meaning of such conditionals is a matter we will discuss in Chapter 5. For now, just be aware that we can't render INUS conditions for the single case in terms of the truth-functional material conditional.

REGULARITIES WITHOUT CAUSATION

The biggest challenge for regularity theories of causation is that not all regularities relate causes to effects. We all have been told that correlation is not causation, and yet a theory of causation grounded in regularities is in danger of asserting just that. To find a way around this difficulty we must first consider the various kinds of non-causal regularities we find in the world, and the generalizations we use to describe them.

Some generalizations are only accidentally true. Suppose that yesterday my meals consisted of a breakfast of croissants and coffee, a cheeseburger for lunch, and cauliflower curry for dinner. Everything I ate yesterday began with "c." This is a coincidence. Nothing about the initial letters of their names caused my menu choices.

There are also many generalizations that aren't just coincidences, but which aren't causal either. Suppose that the sorcerer always hexes his salt before putting it into the pot. It dissolves every time, so we can say that whenever the salt is hexed it dissolves; but it is not the hexing, but the hot water which causes the salt's dissolving. The hexing is causally irrelevant. Here is another example that is less fanciful. All of the currency in my wallet is American, and that is no accident; but putting a bill in my wallet doesn't cause it to become a dollar bill, and no causal mechanisms prevent other bills from getting in my wallet. I have dollars in my wallet because I'm living in the US, but were I to travel abroad, I might have Pesos or Euros in it instead. A generalization like this isn't accidental, but it is true not because it expresses a causal relationship but because of certain background conditions.

An important class of contingent generalizations are generalizations relating two effects of a common cause. For instance, day invariably follows night, but certainly we should not count nights as causing days. The reason is that both day and night (i.e, the rising and setting of the sun) are caused by the earth's rotation. Or consider the relation between falling barometers and storms. Even though falling barometers are regularly followed by storms, falling barometers don't cause storms. Again, what is happening here is that a common cause (a drop in atmospheric pressure) causes both the change in the barometer reading and the storm.

In the examples I've used in this chapter it has been clear what causes what, but much of the time it can be difficult to discern what kind of causal relationship holds between correlated events. Suppose that you are preparing for an exam, and you begin to feel anxious and nauseous. What's going on? One possibility is that you have caught a virus, which is causing the nausea, and your nausea is making you anxious. Another possibility is that your worrying about the upcoming exam is the common cause both of your nausea and your anxiety. Yet another is that the anxiety and nausea are due to different causes.

The situation gets more complex as we move from singular to population level claims, and as we consider more complex and realistic puzzles about causation. Cases of health care disparities provide many examples. For instance, in the first wave of the Covid-19 pandemic in 2020, rates of infection, hospitalization and mortality for Black Americans were significantly higher than for non-Hispanic White Americans. What might have caused this? Race itself might be the cause, either in the unlikely case that there is some significant population level difference in susceptibility based upon some relevant genetic trait, or perhaps because Black Americans receive less quality care in hospitals. It was thought initially that much of the disparity might have to do with underlying medical conditions (like obesity, diabetes, and heart disease) that are, for a variety of reasons, more prevalent in Black communities than in non-Hispanic White communities. What in retrospect seems to be most significant, however, were the differential circumstances in which White and Black Americans lived and worked. Higher proportions of Black Americans were employed in occupations that they could not perform remotely, and Black Americans more often live in more crowded housing that makes social distancing and quarantining more difficult.

An important feature of population level causal claims is that they are in effect averaging over individual cases whose causes may be different. Even if the best causal explanation of the overall disparity in Covid cases between Black and White Americans will appeal to differences in housing and working conditions, certainly many Black Americans live and work in conditions that minimize Covid exposure, and many non-Hispanic White Americans do not.

One other challenge for the regularity theorist is to identify what is cause and what effect when temporal order is unclear. Our ordinary conception of cause supposes that causes precede effects, so when we have clearly identified events, we can always take the earlier event to be the cause of the later event. If the hunter shoots the deer, it is the pulling of the trigger that causes the deer's wound, rather than the other way around. But in many cases where event boundaries are not so clear, or what seems to be a cause or effect is not a discrete event but some ongoing state of affairs, sorting cause from effect is difficult. You are irritated by your roommate's talking too much. But is it really her talking that is making you irritated, or are you irritated with her already for some other reason, which is why you are finding her talking irritating. It's often hard to tell.

DEEP REGULARITY THEORIES

HRT proves to be both too wide and too narrow in what it counts as causes—too wide because it may count an event to be a cause of some effect even if it is an instance of a non-causal regularity, but also too narrow because there can be cause-effect sequences, like those beginning with the big bang, that are not instances of a regularity. Perhaps these problems can be addressed if we take the regularities that ground a causal connection between *c* and *e* not to be a universal association between event types *C* and *E*, but instead to be grounded by a set of deeper laws of nature that really do involve exceptionless regularities.

Here's an example that will illustrate the idea. Hot air balloons behave as they do because of some basic physical laws. When a balloonist heats the air in the bag, the density of that air will decrease. This is a consequence of the ideal gas law ($PV= nRT$), which relates pressure, volume and temperature. Since the bottom of the bag is open, the heating does not raise the pressure in the bag, but instead forces some of the air out of the bag. The difference in density between the air inside and outside the bag creates a buoyant force that, according to Archimedes' principle, is equal to the weight of the amount of the air displaced by the balloon. The balloon will rise when the buoyant force (= weight of displaced air) is greater than the gravitational force upon the balloon (= weight of balloon, including the air within the bag). As the density of the surrounding air decreases with altitude, a balloon maintaining constant temperature air will reach neutral buoyancy and a level altitude. The balloon will begin to sink to earth when the air in the bag cools and the craft becomes negatively buoyant.

Suppose Pierre steps into the gondola and fires up a burner that warms the air in the bag, and that shortly after this the balloon begins to rise. Certainly it is the case that Pierre's firing up the burner causes the balloon to rise. But in virtue of what regularity is this the case? According to a deep regularity theorist, the regularities are the basic laws of nature or physical principles (the ideal gas law, Archimedes' principle, etc.) that, together with a set of facts about the particular circumstances (like the amount of heat generated by the burner, the fact the bag is air-tight, and the weight of the gondola and passengers) entail that the balloon will begin to rise at a certain time and rate.

Contrast this answer to the answer given by HRT. Here the regularity would be something like "whenever you turn on burners in hot air balloons, the balloons rise." This answer is less satisfactory for a number of reasons. First, as we've already suggested, there may not be actual regularities in the world of this kind. Perhaps this is the first and only balloon flight, or at least the first and only flight of this kind of balloon. Second, it may well be the case that sometimes balloons don't manage to rise even though you turned on the burner. This might happen, for instance, if the balloon's bag was not airtight or the gondola was overloaded. The kinds of surface regularities imagined by HRT are typically *ceteris paribus* generalizations at best, since they depend upon conditions being right; but the more basic laws of nature, like Archimedes' principle, are, it is thought, unconditionally true.

Simplifying somewhat a suggestion by David Lewis (1973a), here is a general formulation of a deep regularity theory that I will call DRT. Suppose that C is the proposition that an event *c* occurs, and E is the proposition that an event *e* occurs. Suppose also that L is a set of true statements of laws of nature and F is a set of true statements of matters of fact. Then we can adopt the following truth conditions for causal claims:

Deep Regularity Theory (DRT): *c* causes *e* if, and only if,

1. C and E are true (i.e., *c and e* occur)
2. L plus F entail $C \supset E$.
3. L plus F do not entail E.

The point of clauses 2 and 3 is to connect the occurrence of *c* and *e* to laws of nature. Clause 2 implies that, given the laws and facts, if *c* occurs, then *e* occurs; while clause 3 implies that without *c*, the laws and facts are not sufficient to entail that *e* occurs. In our given example, given the facts and laws of nature, Pierre's turning on the burner will be followed by the balloon's rising; but absent his turning on the burner, the balloon will not rise.

There is much to like about deep regularity theories. They explain how cause-effect relations ultimately depend upon lawful regularities, while allowing that many higher-level causal associations, like the association between striking matches and lighting them, or the association between smoking and cancer, will admit of exceptions. They are Newtonian in spirit, in the sense that they seek to show how a wide variety of phenomena may be accounted for by a few basic laws.

DRT is not though without its difficulties. For one thing, it is not an epistemologically practical account, in the sense that it doesn't really tell you how to decide whether particular happenings in the world are causally connected. When I smile at my neighbor and they smile back, I think I have reason to believe that my smile caused their response, even though I have no idea about how these events are connected through fundamental laws of nature. More importantly, DRT is not complete without some account of laws of nature that makes clear which regularities are lawful and which are not. The question of what laws of nature are has been the subject of philosophical discussion for centuries and continues to this day. We'll consider some basic issues and positions below.

LAWS OF NATURE

Philosophers do not agree about the relation between causes and laws of nature. A common view, of which the deep regularity theory is an instance, supposes that relations of cause and effect depend upon laws. But some, like Russell, have claimed that causes can be dispensed with and replaced by laws. Still others have claimed that a cause can bring about an effect in the absence of law. Regardless of what view is ultimately correct, it is clear that many things we call laws of nature appear to describe or underlie causal relationships, so we need to consider some different accounts of what laws of nature are.

Let's begin with examples. We've already mentioned a number: Newton's law of universal gravitation, Archimedes' principle, and the ideal gas law. These examples are all drawn from physics, where law talk is most common, but there may be examples from other sciences—perhaps Mendel's laws in genetics, or Darwin's principle of natural selection, or maybe the law of effect in psychology or the laws of supply and demand in economics. Beyond the laws found in textbooks, there are many other generalizations that appear to express natural laws. For instance, it may well count as a law that all metals conduct electricity or that every human needs oxygen to survive.

To think carefully about what laws are, it will be helpful to distinguish between laws and law statements. Law statements are bits of language or mathematics, including expressions like "$PV=nRT$" or "$F_G=Gm_1m_2/r^2$." In contrast, laws themselves are features of the natural world—things that do not depend upon us or our theories, but which we hope to discover through scientific research. Newton, for instance, discovered the law of universal gravitation, expressed by the law statement, "$F_G=Gm_1m_2/r^2$." Laws themselves are whatever it is in the world that make law statements true. Commonly we use the word "law" to refer both to law statements and to the actual laws which are the truthmakers for those statements. This is harmless enough, so long as we understand that when we are asking "What is a law of nature?" we are asking a question not about statements, but about the worldly facts that make those statements true.

Newton's law of universal gravitation, as well as many of the other examples I've mentioned, are expressed by equations that describe relations between quantities. These are often called functional laws, and are the kind we most commonly see in science textbooks. Philosophers of science though have tended to focus on a simpler kind of law, the kind which can be expressed by universal generalizations relating kinds or properties. For instance, it seems like a law that all humans are mortal, or that all metals conduct electricity. Philosophers like these examples, because they provide simple cases which make it easier to bring out essential metaphysical questions about the nature of laws.

I've used the terms "generalization" and "universal generalization" already, but because their structure is important to our discussion of laws of nature, let me say more explicitly what these things are. Let's begin again with examples:

All children have parents
Some Germans speak Spanish
Most Australians speak English
More than 90% of people who get lung cancer are current or former smokers

The first statement is a **universal generalization**. The general form of such a statements is "All *F*s are *G*s"—in predicate logic "$\forall x(Fx \supset Fx)$"—and it will be true so long as everything that is an *F* is also a *G*. The second statement is an **existential generalization** whose form can be expressed as "Some *F*s are *G*s"—in predicate logic "$\exists x(Fx \& Gx)$." Since "some" is understood to mean "at least one," such statements will be true so long as there is at least one thing that is both *F* and *G*. The third and fourth statements are quantitative/statistical generalizations, which we will discuss in the next chapter. All of these generalizations describe patterns, associations, or regularities in the world. Our question is, which such generalizations might describe laws of nature?

Whatever they are, laws of nature must exhibit some kind of necessity. Regularity approaches to laws of nature interpret necessity as universality; they take "must" to mean "always." For this reason, regularity theorists assume that natural laws must be universal generalizations of some kind. But, as we've already seen, there are lots of true universal generalizations that don't seem to express laws, because they are only accidentally true. If everybody in my class is under six feet tall, that would simply be a coincidence. And if all of the bills in my pocket are US dollars, that's not a coincidence—I'm living in the US after all—but neither is it a law, since nothing about bills being in my pocket makes them US dollars.

One common way to characterize the necessity required of laws of nature is to say that laws must support counterfactuals. Not only must all the *F*s be *G*s, but it must be the case that if something were *F*, it would also be *G*. The reason that it's a law that all metals conduct electricity is that, were something nonmetallic to be made of metal, it would conduct electricity. The reason it's not a law that all the students in my class are under six feet tall is that were another student to be added, they wouldn't necessarily be under six feet tall. But while the ability to support counterfactuals is clearly a marker of laws, this won't help us understand what laws truly are unless we get clear on what makes counterfactuals true; and that, we'll see in Chapter 5, is a tricky problem. Moreover, many people think that the truth of counterfactuals depends upon the truth of laws, and if this is the case, we can hardly use the criterion of counterfactual support to distinguish laws from non-laws.

The regularity theorist needs to somehow add some criteria to distinguish generalizations that express laws from true but non-lawful generalizations. Logical empiricist accounts of laws typically use criteria like these:

1. The generalization must not be a logical or conceptual truth
2. The generalization must contain only predicates that are not spatio-temporally restricted

The first criterion rules out generalizations like "All triangles have three sides" or "all bachelors are unmarried." These statements follow from the meanings of words; they are, as Hume would say, not concerned with matters of fact, but relations of ideas. The second criterion is meant to bar generalizations that are only accidentally true. A spatio-temporally restricted predicate is some expression that applies only to particular individuals, or to particular places and times. Phrases like "is in my classroom" or "is American" are spatio-temporally restricted. Unrestricted predicates are things like "has a mass of 10 kg," "is negatively charged," or "is a gas." The rationale for this criterion is that the unrestricted predicates refer to intrinsic properties of individuals, while the restricted ones don't express properties of the individual itself, but relations between the individual and its context. It's true of me right now both that I weigh 190 pounds and I am living in Indiana. The first claim is about a genuine property of me (my weight), but the second is not, because you could change my location without changing anything about me. (Actually, if you consider the matter more closely, weight isn't intrinsic either; you can change my weight without changing me, simply by moving me to the moon. A better candidate for a genuinely intrinsic property (at least within classical physics) is mass, since you can't change an individual's mass simply by changing its position with respect to other individuals.

These criteria may be enough to rule out a lot of accidental generalizations, but there are some generalizations that appear to satisfy them, but which still are not laws. One problem is laws without instances—also called vacuous laws. The rules of logic imply that if there are no *F*s, the generalization that all *F*s are *G*s will be true for any *G* whatsoever. It will be true, for instance, that all flying elephants have a maximum airspeed less than 100 miles per hour, simply because there are no flying elephants. Still, we wouldn't want to call this generalization a law.

Another problem is what we might call cosmic accidents. It may well be the case that all spherical chunks of gold in the universe have a mass of less than 1,000 kg, but this is hardly a law. It just happens, given the distribution of matter in the universe, that all the gold spheres there are smaller than this. On the other hand, it probably is a law that all spheres of uranium 235 have a mass less than 1,000 kg. The reason is that, given principles of radioactive decay, it's not possible to have such a large sphere of U-235 without it exploding.

These and other difficulties have led philosophers to look for a better account of laws of nature. Two approaches have received the most discussion, and the differences between them will be helpful in thinking about how laws relate to causes. The first approach, initially suggested by Mill, and developed by Frank Ramsey and David Lewis, is Humean in spirit, since it attempts ultimately to reduce the necessity of laws of nature to regularities. The second, developed by David Armstrong, Fred Dretske and Michael Tooley, is anti-Humean because it supposes that laws of nature depend upon an irreducible relation of necessitation.

The Mill-Ramsey-Lewis (hereafter MRL) approach is often called the best-systems account of laws of nature. The approach draws its inspiration from axiomatic theories in mathematics. In an axiomatic theory, one identifies a generally small set of axioms or postulates, from which other propositions, the theorems, can be derived by deductive inference. The first well-developed axiom system was given by Euclid in his *Elements* more than 2,000 years ago, but in the 19th and 20th centuries, axiom systems were developed to describe other mathematical theories (like arithmetic) as well as physical theories (like Newtonian mechanics). The MRL approach suggests that the fundamental laws of nature can be described by axioms of such a system, and theorems expressing less fundamental laws can be derived from these axioms. Imagine, for instance, that Newton's law of universal gravitation, along with his laws of motion are fundamental laws. Other laws, like the law of falling bodies or Kepler's laws of planetary motion, would be theorems derived from the basic laws.

But the question is, what axioms should we choose, and how do we know that they express true laws? Lewis (1973b) argued that we should choose axioms which provide the best trade-off between simplicity and strength. To get an idea of what simplicity and strength might mean, consider again the history of astronomy. The Ptolemaic system of astronomy, while it had its inaccuracies, had been tuned by the Middle Ages with more than 1,000 years of data and made lots of accurate predictions about the location of stars, sun, moon and planets. But to make it explain the peculiarities of planetary motion, the system had to contain lots of complexities. The planets did not simply orbit the earth; Ptolemy needed to add other devices—called epicycles, deferents, eccentrics, and equants—to make the predictions come out right. Copernicus's system, at least in its basic form, looks much simpler. The earth is a planet, and, like all planets, orbits the sun in circular motions. This system got the basics of planetary motion right, providing a simple explanation of why Venus never strays far from the sun, or why Mars occasionally moves backward. Still, without adding additional complexities, Copernicus' system was less accurate than Ptolemy's. So it looks like a Copernican system was simpler but, at least initially, the Ptolemaic system was stronger. Sun-centered astronomy improved with Kepler's suggestion that the planets follow elliptical orbits with speeds that vary with their distance from the sun. Kepler's improvements allowed the ditching of a lot of the messy math associated with Ptolemy and Copernicus, and led to a system that was both simpler and stronger than its predecessors. Newton's system was in turn simpler and stronger than Kepler's, deriving and correcting Kepler's laws, while unifying celestial and terrestrial motion.

Newton's system was still not the best, and simpler and stronger systems have been developed subsequently (most notably with the introduction of Einstein's general theory of relativity). No one would assert that even now we have found the best system to describe all of the natural phenomena we see in this world, so we almost certainly don't know what the most basic laws of nature will be, but the idea is that as science progresses, we get closer and closer to finding them.

A major virtue of the MRL approach is that the laws it identifies generally coincide with what we take to be intuitively plausible candidates for laws of nature. For instance, we think the principles of Newtonian mechanics might be laws, because they provide a simple and economical description of a lot of the goings on in the world. But there are two big problems with the MRL approach. The first is that it seems to give a subjective answer to a question for which we seek an objective answer. There's no real objective principle by which we can judge the proper balance of simplicity and strength. It's a bit like judging the proper balance between speed and fuel efficiency in a car—it's all about what you care about. Additionally, there's no clear metric for deciding when one system is simpler than the other. Simplicity may be in the eye of the beholder.

The second problem is that the MRL approach does not explain *why* things happen as they do. It simply provides us the strongest and simplest description of what Lewis has called the Humean mosaic. That mosaic is just the sum total of all the happenings in the world, from supernovae in distant galaxies, to the flightpath of a hawk in the sky above you, to the footprint you left in the sand. Because the description is simple, it will detect many patterns and regularities in the mosaic. What it won't tell you is why those patterns are there. This may be acceptable to a Humean, who believes that the mosaic is all there is. But if you are looking for an account of laws that explains the patterns rather than just describing them, you will need to look beyond the MRL account.

According to advocates of the Armstrong-Dretske-Tooley (ADT) account, the mistake of the MRL and other Humean accounts is that they take laws to be grounded in the patterns of particulars in the Humean mosaic. Take again the supposed law that all metals conduct electricity. On a regularity account, what makes this true is just the fact that every particular thing in the world that is metal also is a conductor. The law really just is the pattern of particulars. MRL theorists argue that these patterns need explanation, and what explains them is the law.

What then is the law? MRL theorists claim it is a necessitation relation that holds between properties. For the law that all metals conduct electricity, the necessity holds between the property of being a metal and the property of being a conductor. All particular pieces of metal are conductors because being a metal necessitates being a conductor. To put the matter generally, when we say that it is a law that all *F*s are *G*s, we are really asserting that *F*-ness necessitates *G*-ness. In his version of the theory, Armstrong (1983) denotes this relation as "*N* (*F*, *G*)."

To clarify the ADT view, we need to say more both about what things stand in the *N* relation and about what the *N* relation is. Armstrong claims that the *N* relations holds between **universals**. "Universal" is an umbrella term encompassing (1) properties of things, like color, emotion, or charge; (2) kinds of things, like a metal, a gas, or a vertebrate; and (3) relations between multiple things, like the distance between two objects or the biological relation between mother and child. Universals are distinguished from particulars, which are the individuals that can have properties, be of a kind, or stand in a relation. To see why Armstrong needs universals in general rather than just properties, consider the universals involved in the law of universal gravitation. On the ADT reading, that law expresses how the force acting between two objects (a relation) is necessitated by their masses (properties) and their distance (a relation).

To help clarify what this relationship *N* is, and why it is a relation between properties and other universals, let me borrow an analogy from Dretske (1977). Dretske argues that the natural necessitation that holds between universals is analogous to the legal necessitation we see in a constitutional government. We know, within the US constitutional system that there are different offices with different powers and duties. The president, for instance, can nominate Supreme Court justices, and no one else can. The senate in turn must approve the nomination, and only then can the new justice be sworn in. Particular presidents nominate particular people—Obama nominated Kagan, Trump nominated Gorsuch, Biden nominated Jackson—but they can only do so when they hold the office and in virtue the office they hold. Similarly, the senators who approve the appointment can do so only when and because they hold the office of senator. The constitution sets the powers of the offices, and it is these powers which explain what the office holders can and must do. In laws of nature, the properties and other universals are like the offices, and the collection of necessitation relations holding between them is like the constitution, which governs how things with these properties behave.

The analogy also explains the sense in which natural laws are *contingent necessities*. Given the constitution, the duties and powers are fixed—they are necessities. But the constitution itself is contingent. It could have been different and it could be changed. Similarly, given that the relation *N* holds between various *F*s and *G*s, the necessary connection between *F*-ness and *G*-ness is explained. But it is possible to imagine that the laws of nature might be different, just as the constitution might be changed.

Critics of the ADT approach to laws still find this necessitation relation mysterious. One does not just by giving it the name "*N*" tell you what it is. When pressed on this subject Armstrong replied that the relation is simply causation (Armstrong 1993). Maybe this is right, but it is troubling in two respects. First, there are some laws, like conservation laws in physics, which don't seem to relate causes and effects. Second, if we are going to appeal to causal relations to explain what laws are, we are going to have a hard time using laws to explain what causes are.

HUMEAN ANSWERS TO THE QUESTIONS ABOUT CAUSATION

As we close this chapter, let's consider how regularity theorists have answered the questions about causation we identified in Chapter 1. Humean approaches to causation answer conceptual and metaphysical questions, but their starting point is with epistemological questions about how we could come to know causes and effects. These approaches accept Hume's empiricism—the assumption that all knowledge derives from experience. They also accept Hume's argument that causal connections cannot be observed, and that all we can actually observe are regular patterns of association. From there it is not a long step to adopt a metaphysical position that causation is nothing but these regularities. The position needs refinement since not all regularities are causal, and the task of the regularity theorist is to give some account of which of them are.

Regularity theorists of all stripes reduce causal connections to patterns of difference-making. At its simplest, the claim that *X*s causes *Y*s reduces to the claim that whenever *X*s occur *Y*s follow, and this is just a matter of *X* making a difference to *Y*. This simple version is subject to counterexamples, leading philosophers to offer more nuanced versions of difference-making, like those found in INUS conditions or in the laws of nature invoked by deep regularity theories. Still, all of these regularities are ultimately just patterns of association between things that happen—the Humean mosaic. Regularity theories give accounts of the truth conditions for both singular and generic causal claims, but generic causal claims drive the account, because what ultimately makes a singular causal claim true is that is an instance of a generic causal pattern.

Another question to consider is whether the concept of cause analyzed by regularity theories is a commonsense concept, or a scientific one. The answer depends upon the theory in question. The HRT and INUS condition accounts are centered on commonsense conceptual considerations: what do people mean when they assert things like "Falling into lakes causes drowning" or "fires cause warming"? On the other hand, deep regularity theories of causation and the MRL approach to laws of nature are more scientific in spirit, as they see causal relations as depending upon some set of basic laws, interpreted as regularities, of the sort that scientists seek to uncover.

With the exception of the ADT theory of laws, the theories we have considered in this chapter are both conceptually and ontologically reductive. Conceptually they seek to define causes and effects in terms of more basic concepts, and ontologically, they reduce the supposed necessary connection between cause and effect to some pattern of constant conjunction.

The ADT theory is an important exception, and the first pushback we've considered against the Humean approach. For ADT theorists, regularities cannot explain what things are lawfully and causally connected. Instead, these regularities are what laws or causes need to explain. The ADT theorists take necessity as something fundamental which accounts for the patterns in the Humean mosaic. For this reason, they are generally accounted as realists about laws, and by extension, about causes.

Our last question asked whether a theory of causation treats what causes what as a matter of fact or a matter of convention. Regularity theories generally fall on the side of fact, since whether or not some potential cause and effect are instances of an appropriate regularity is a matter of the way the world is. But there is one place where conventional issues may creep in. The MRL approach to laws (which on deep regularity approaches ground causes) identifies laws as axioms of best systems, but what makes for the best system depends upon a balancing simplicity and strength—and how one chooses to balance simplicity and strength may not itself be forced by the way the world is, and may reflect preferences or conventions.

SUGGESTIONS FOR FURTHER READING

Psillos (2009) provides an excellent overview of contemporary discussions of regularity theories, including a helpful discussion of the connection between regularity theories and laws of nature. Mackie's (1980) *Cement of the Universe* contains a classic exposition of the regularity theory in Humean terms. Mackie's "Causes and Conditions" (1965) contains the original account of INUS conditions. Classic anti-Humean accounts of laws of nature can be found in Armstrong's (1983) *What is a law of nature?* as well as in (Dretske 1977).

4

CAUSATION, PROBABILITY, AND CHANCE

The regularity theories of causation discussed in Chapter 3 took causes to be sufficient for their effects. In the case of a simple regularity account, *c* causes *e* just in case *C*s are always followed by *E*s. The INUS condition account improved on the regularity account by recognizing that things we call causes are often not sufficient for their effect, but is instead part of a conjunction of causes which are jointly sufficient.

Sufficiency analyses, whatever their virtues, seem ill equipped to explain another familiar kind of cause—a cause that makes its effect more likely without guaranteeing it. A great deal of the causal knowledge we have is about things that raise or lower the chance of something happening. If a disease is infectious, we know that contact with a disease bearer makes it possible or likely someone will contract the disease, but it doesn't make it certain. Similarly, masks or vaccines may reduce the chance of contracting diseases, but they don't generally guarantee one won't get sick. Smoking, we all know, causes lung cancer, but not all smokers contract lung cancer, and some non-smokers do.

Our goal in this chapter will be to explore the meaning of such causal claims, and to see whether we can give an account of causation grounded in the idea that causes raise the probabilities of their effects. Probabilistic theories of causality are difference-making theories, and most can be understood as elaborations of regularity theories, but the regularities they track are statistical rather than deterministic.

DOI: 10.4324/9781003200956-4

Before introducing these theories, it will be helpful to recall a few facts about probability. Think of a probability as a function which assigns to any event a number between 0 and 1. What we take these events to be will depend upon what we are trying to use the probabilities to represent, but for our purposes, we can think of them as occurrences you might observe—like a flipped coin showing heads, or you contracting a virus. You can express the fact that a coin flip has a 50% chance of being heads like this:

$$P(H) = .5$$

We can use logical operations to create compound events and assign probabilities (using "~" for "not," "V" for "or," "&" for "and"). So if A and B are events, we can assign probabilities to compound events, e.g., $P(A\&B)$, $P(\sim B)$, or $P(\sim A \vee \sim B)$. When events A and B are mutually exclusive, i.e., $P(A\&B) = 0$, you can get the probability that either will happen simply by adding the probabilities that each will happen. For instance, drawing a diamond from a full deck and drawing a spade from a full deck are mutually exclusive, so $P(D) = .25$, $P(S) = .25$ and $P(D \vee S) = .5$. When events are independent of each other, in the sense that the occurrence of one does not make a difference to the occurrence of another, you can multiply probabilities. For instance, if you flip a coin twice your probability of getting heads both times is $P(H_1\&H_2) = P(H_1)$ $P(H_2) = .25$.

Crucial to the development of a probabilistic theory of causality is the idea of conditional probability. A conditional probability expresses the probability of one event occurring given another has occurred. Conditional probability is symbolized as $P(A \mid B)$ (read "the probability of A given B"), and can be defined as follows:

Conditional Probability: $P(A|B) = P(A\&B) / P(B)$

For instance, continuing the coin flip example, $P(H_1\&H_2 \mid H_1) = .5$.

WHAT DO PROBABILITIES MEAN?

The mathematical apparatus of probability theory is open to multiple interpretations. The most basic question an interpretation of probability must answer is whether probabilities represent objective facts about the world, or subjective facts about our knowledge and beliefs. Suppose that you have shuffled a deck of cards and are about to draw the top card off the deck. What's the probability that it's an ace? The obvious answer is that it is 1/13. But if you reflect further, you may want to reconsider. Why? Because the top card is already determined by the shuffle. If there's an ace on the top of the deck, your chance of drawing an ace is 1, while if it isn't there, your chance is 0. The fact that you were inclined to say the chance was 1/13 is simply a reflection of your ignorance. Objectively there is no chance at work at all.

There are two overlapping ways to think about subjective probability. One, which we can call evidential probability, is an epistemic measure—an estimate of the likelihood of something given the available evidence. The other, which we can call personal probability, is a measure of an individual's degree of belief. The concepts generally align, because, when we are acting rationally, our degrees of belief should be determined by the evidence we have. Consequently, while these kinds of probabilities are subjective in the sense of being about knowledge and belief rather than features of the world, they need not be arbitrary or irrational. In our example above, for instance, the answer of 1/13 is the correct probability you should assign to the Ace lying on the top of the deck, given the evidence you have.

If probabilistic causes are features of the world, however, we will need though to look for an objective interpretation of probability. The most common such interpretation is called the frequency interpretation. In this interpretation, probabilities describe the relative frequency of an occurrence given some kind of "random setup." For instance, to say that the probability that a coin coming up heads is 50% is simply to say that in the long run the relative frequency of heads being the result of a coin toss will approach 50%. Flip a coin five times, and you might get 80% heads, or 0%, but the longer you toss coins, the closer you'll come to a rate of 50%. The frequency interpretation is a natural way of characterizing probabilities of events that occur to a certain proportion of individuals in a population. When doctors state the risks of complications in pregnancies or surgeries, these are probabilities that represent relative frequencies.

Conditional probability statements, understood as expressing relative frequencies, can be used to make generic causal claims. So, for instance, the causal claim that smoking causes cancer can be interpreted as the claim that the relative frequency of cancer among smokers exceeds that of non-smokers.

It is much less clear how we should interpret probabilities used in singular causal claims. Suppose that I carry a gene, and that 10% of people carrying this gene develop colon cancer, while only .1% of the population as a whole does. Does this mean that I really have a 10% chance of contracting cancer? Probably not. My chance of getting cancer will depend not just upon this gene, but upon other genes I carry, as well as other facts about my past and current environment and lifestyle. If we took into account all of these factors, these might be jointly sufficient either to put me in the 10% of gene carriers who get cancer, or to keep me out. When I think of myself as having a 10% chance of getting cancer, this might simply reflect my ignorance of the role of these other factors, getting us back to subjective probability.

Often what lies behind this intuition that there are not really any single case objective probabilities is a commitment to determinism—the idea that the totality of facts at a given time together with the laws of nature determine the totality of facts at all later times. We cannot dive deeply into whether this commitment is justified, but we should briefly at least address two questions. First, is there reason to believe that determinism is true? And second, is there any way to make sense of objective single case probabilities if it is not? The first question should be informed by our understanding of physical theory. Determinism was a widely held view in the 18th and 19th centuries, largely because of the remarkable success of Newtonian mechanics, which allowed for the precise prediction of many celestial and terrestrial phenomena. It was not a long step to imagine that a complete physics would yield a set of laws that showed how all future states of the universe were determined by its current state. Probabilistic causal claims could remain useful, because we so often don't know all of the facts which determine events, whether they be coin flips or contracting a disease, but these would be subjective probabilities—reflections of our ignorance.

This scientific story took an unexpected turn in the early twentieth century with the discovery of quantum mechanics. Quantum theory, which describes the behavior of entities at atomic and subatomic scales, suggests that certain events are irreducibly indeterministic. Pass a beam of electrons through a magnetic field, and they will scatter in a predictable pattern, but—for reasons we can't consider here—there appears to be no underlying fact about why each individual electron goes the direction it does. Given that everything in the world is ultimately composed of atoms and subatomic particles, it is natural to think that this indeterminism could percolate upward into the behavior of larger things—cells, trees, baseballs, asteroids, and so on. If this is what the world is like, objective probabilities could describe propensities, indeterministic tendencies of things towards certain outcomes. For instance, an electron going through a magnetic field would have a propensity that would determine the likelihood of it scattering in various possible directions.

My quick sketch of what physics has to tell us about determinism leaves a lot out. Classical mechanics can be indeterministic in certain circumstances, and quantum mechanics can be given a deterministic interpretation. But most importantly, nothing in this sketch should give us reason to believe that the probabilities we use to describe the behavior of objects above the atomic scale represent quantum indeterminacy. Even if quantum mechanics is indeterministic, for instance, there is little reason to believe that quantum phenomena are necessary to explain why the coin flips heads this time and tails the next.

The second question we needed to consider is whether we can make sense of single case probabilities where determinism is true. A few examples will suggest a way. Consider first the behavior of a Roulette wheel. Suppose determinism is true and that in fact given the very precise facts about how the wheel is spun and where the ball started, there is a definite slot where the ball will land. Is there still a reason to think of the ball, before it spins, as having an equal chance to land in each of the thirty-seven slots? Perhaps so. Roulette wheels are designed as randomizing devices. There are not (or at least should not be!) any hidden mechanisms to bias the outcomes, and ever so slightly different spins lead to very different results. As a second example, consider a moth who feeds on trunks of trees. Among the moths there are different patterns of camouflage that are differentially able to hide them from the eyes of predators. We might imagine that the better camouflaged moth has an 90% chance of surviving the day, while the less well camouflaged one has an 80% chance. Can we attach different propensities to these two moths? A determinist might argue not. After all, if we knew the exact flight path of every predator flying by that day, and knew everything about their vision and their cognition, we should be able to predict which of the two moths will survive. Nonetheless, we might respond that the different camouflage patterns represent real facts about the two moths—facts that make them differentially visible to predators, and differentially likely to survive. These propensities in turn explain (via natural selection) why one pattern comes to predominate. These propensities seem to express some fact about the world's causal structure. Notice how different these two cases are from the subjective chance in the cancer case, where there may well be facts about me that I don't know but which will determine whether I get cancer or not.

PROBABILISTIC CAUSALITY

Probabilistic theories of causation are theories that use the tools of probability theory to represent causal structure. There are quite a number of such theories, but in this chapter, I'll sketch a basic version of the approach that originates in the work of Hans Reichenbach (1956). The intuition that motivates most probabilistic theories of causation is that causes should make their effects more likely. As a first pass, consider a simple theory—we'll call it PC_1—that uses this criterion to define causality as a relation between two generic types of events:

Probabilistic Causation, (PC_1) : *A* causes *B* if, and only if, $P(B|A) > P(B)$

Applying this to our example of smoking and cancer, PC_1 implies smoking causes cancer just in case the probability of cancer is greater among smokers than the population at large. (Some philosophers express this principle by saying that the probability of an effect given a cause is greater than the probability of an effect given the absence of that cause, i.e., $P(B|A) > P(B|{\sim}A)$, but this turns out to be equivalent to PC_1.)

PC_1 offers some immediate benefits over non-probabilistic regularity theories. By replacing exceptionless regularities with probabilistic regularities, the probabilistic account allows for causal generalizations that have exceptions. If you flip a switch in your house the light may not go on (because the power is out, the bulb is burned out, etc.), but certainly it raises the chances. The probabilistic theory also helps address the causal irrelevance problem. Recall the hexed salt problem from the last chapter. Even if the hexed salt always dissolves, the hexing of the salt doesn't raise the probability of the dissolving, so hexing doesn't count as a cause.

The chief problem with PC_1 is that it is too permissive; it counts any positive correlation between event types as a causal relation. It would count falling barometers as causing storms and buying cribs as causing babies. PC_1 will thus have to be supplemented with criteria to pick out which correlations are indicative of relations between a cause and effect. If generic events *A* and *B* are correlated, i.e., $P(A\&B) > P(A)P(B)$, it is safe to assume that there is some causal relationship that explains this correlation, but the correlation by itself doesn't tell you what that causal relationship is.

Three basic possibilities are depicted in Figure 1. Either (a) *A* causes *B*, (b) *B* causes *A*, or (c) *A* and *B* are caused by some common cause *C*. If the temporal relationship between events is clear, we can use that order to decide the direction of causes. We know for instance that sex causes babies, rather than babies causing sex, since babies only come after sex. But the fact that event *A* precedes and is correlated with *B* is still not enough to guarantee that *A* causes *B*, since *A* and *B* may be correlated via a common cause. To distinguish cases (a) and (c) we must use another criterion called screening off ex:

Screening Off: *C* screens off *A* from *B* if, and only if, $P(B|C\&A) = P(B|C)$

If *C* screens off *A* from *B*, once the probability changing effect of *C* on *B* has been taken into account, *A* makes no further difference to *B*. In the case of barometers and storms, pressure drops (*C*) screen off barometers falling (*A*) from storms coming (*B*).

Figure 1d pictures a second situation in which *C* screens off *A* from *B*, namely a situation where *C* is a causal intermediate between *A* and *B*. As an example, imagine some kids playing baseball in the back yard. One of them swings at the pitch (*A*), striking the ball (*C*) and sending it towards the window which it breaks (*B*) Swinging the bat raises the probability of breaking the window, $P(B|A) > P(B)$, but bat swinging is screened off by ball striking, $P(B|C) = P(B|C \& A)$. Given that the ball is struck, the fact that the bat was swung has no additional effect on the chance of breaking the window.

A theory of probabilistic causation will want to rule out *A* as a cause of *B*, if *A* and *B* are correlated via a common cause (1c), but it should count *A* as a cause of *B*, even if there are intermediates between *A* and *B* that screen off *A* (1d). We can use time-ordering of events to distinguish these cases, and give us an improved account of probabilistic causation.

Probabilistic Causation$_2$ (PC$_2$): *A* causes *B* if, and only if:

1) $P(B|A) > P(B)$
2) *A* occurs before *B*
3) There is no *C* that occurs before *A*, which screens off *A* from *B*.

Condition 1 asserts that *A* raises the probability of *B*; condition 2 uses time order to make sure that *A* causes *B* rather than *B* causing *A*; and condition 3 makes sure there is no common cause *C* that occurs prior to *A* & *B*, which explains the correlation between *A* and *B*. If *A* causes *B* via an intermediate (as depicted in Figure 1d), it still counts as a cause according to PC_2.

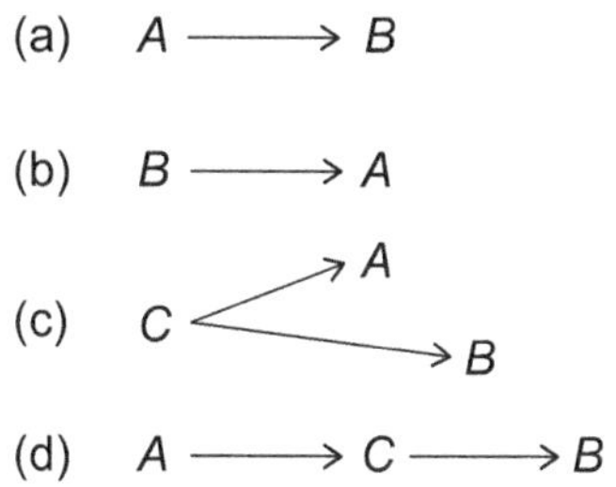

FIGURE 1. Causal structures that create correlations

A probabilistic theory of causation along the lines of PC_2 makes considerable progress towards understanding the relationship between correlation and causation. It recognizes that positive correlations are signs of causation, but uses conditions 2 and 3 to rule out non-causal correlations.

While from a theoretical point of view, this account can distinguish probabilistic causes from non-causal correlations, it may in practice be very difficult to get the information required to decide whether a correlation is causal. We'll discuss this matter in more detail in Chapter 9, but for the moment let's consider a brief example. Suppose that within a population of high school students you have discovered correlations between classroom misconduct, truancy and standardized test performance. Which of these behaviors might be causing which? A first major difficult is there is no clear causal order between these variables, since they are not particular dated events, but rather patterns that overlap in time. A related problem is that behaviors like this may be accounted for by unmeasured common causes. For instance, unstable family circumstances or a learning disability might influence each of these variables. Perhaps most importantly, there is no reason to believe that the causal relationships between these variables will be the same for each student. What explains one student's poor test scores need not be the same as what explains those of another student. If this is the case, then it may be impossible to say anything general about the causal relationship between these variables.

SIMPSON'S PARADOX

Probabilistic theories of causation can founder because of a statistical phenomenon called Simpson's paradox. The basic difficulty is that if A is a *prima facie* cause of B, i.e., $P(B|A) > P(B)$, it is generally possible to partition the population of Bs into subpopulations, where in each of these subpopulations, the effect of A is neutral or even negative. I'll illustrate this phenomenon by reference to a real-life case discussed by Nancy Cartwright (1979). In the 1970s it was observed that men were being admitted to the Berkeley graduate school at higher rates than women, and an investigation was conducted to determine whether this was so. What the investigators found was that if admissions were partitioned by department, the effect disappeared or reversed. To illustrate this effect, let's consider a simple hypothetical set of admissions data:

		Asian Studies	*Biology*	*Chemistry*	*English*	*Total*
Men	Applied	20	50	100	50	220
	Admitted	5	18	55	25	103
	Rate	25%	36%	55%	50%	47%
Women	Applied	10	80	15	25	130
	Admitted	3	30	9	13	55
	Rate	30%	38%	60%	52%	42%
Overall	Applied	30	130	115	75	350
	Admitted	8	48	64	38	158
	Rate	27%	37%	56%	51%	45%

In each of the four departments, women are admitted at higher rates than men. Nonetheless, men are admitted at a higher rate overall, because the departments to which the men applied in higher proportions (Chemistry and English) have higher overall admission rates than the departments where women applied in higher proportions (Asian Studies and Biology). Among 85 graduate programs at Berkeley, investigators found something very much like this effect.

Partitioned by department, this data suggests that being a woman helps applicants rather than hurts them. Would such data show that it was actually the men who were suffering discrimination? Not necessarily. For one thing there are many other causally relevant factors that would have to be considered. Admissions should be affected by applicant's abilities and accomplishments as measured by things like test scores, GPA, quality of undergraduate program, and so on. Perhaps women's higher rates of achievement could be explained by their scoring on average higher than male applicants on these criteria. Such criteria could screen off gender effects to a point where, conditionalized on these criteria, men and women perform equally well. It's also possible that, conditionalized on these criteria, that apparent negative effects of being female might reappear at the department level. And there are many other potential sources of discriminatory effects. For instance, it could be that the criteria used for admission (test scores, assessments of institution quality, etc.) are systematically biased against women, and one might wonder, given the disparity in number of men and women applying, if there were gender-based obstacles to application.

Simpson's paradox reminds us that when we see a positively statistically relevant factor in a population, we cannot immediately infer that it is causally relevant. We must instead look for partitions of the population, and see if the statistical relevance remains. But how do we do this? The first step is to identify all of the factors that are causally relevant to the effect. In our example, this might include test scores, GPA, race, age, and many other things. These factors, taken in combination, constitute the *background context* in which *C* acts on *E*. Given these background contexts, Cartwright suggests the following definition of a cause:

> **Contextual Unanimity:** *C* causes *E* if, and only if, for all background contexts B_i $P(E|C\&B_i) > P(E|B_i)$

This is called a contextual unanimity requirement, because on this criterion, the cause must raise the probability of its effect in *all* background contexts.

One big problem with this criterion is deciding what count as legitimate background contexts. There are many ways we could partition the population that would not be causally relevant. For instance, we could divide applicants by birth months, or by the last digit of their social security number. Students with February birthdays might as a matter of fact get in at higher rates than students with March birthdays, but we'd be extremely unlikely to think that that was anything but a matter of luck.

The problem is that we need to rely on prior causal knowledge to decide which partitions are or might be causally relevant.

Even if we could figure this problem out, we might wonder whether contextual unanimity is simply too strong a condition. Consider a hypothetical situation in which men were admitted at higher rates in 82 out of 85 departments, but women had an advantage in the other three. According to contextual unanimity requirement, being male wouldn't count as a probabilistic cause of getting into grad school. But in such a case we surely would want to say that being a woman hurts applicants, because in most cases it does. For this reason, some philosophers have proposed weakening contextual unanimity, so that *C* is a cause of *E* if it raises the chance of *E* on average, or most of the time. While it seems clear that the contextual unanimity requirement does not fit with our ordinary use of causal concepts, neither does it seem like we are forced to accept or reject the principle on scientific or philosophical grounds. This seems like a case where the decision to adopt a standard is a practical one—a matter of choosing an appropriate convention.

SINGULAR CAUSATION AND DOING IT THE HARD WAY

Probabilistic accounts of causation are most naturally suited to interpreting generic causal claims about populations. If smoking causes cancer, on the probabilistic account this means roughly that smoking raises the probability of getting cancer within human populations. Similarly, in the discrimination example, the claim under consideration was whether being a woman lowered the probability of admissions within the population of graduate applicants to a particular school.

But, as we noted already in our discussion of interpretations of probability, sometimes it makes sense to make probabilistic singular causal claims about individuals. For example, I might advise a student who wants to get into law school to retake the LSAT because I think raising her score will increase her chance of admission. Suppose she does retake the test and raises her score. If she then gets into a law school that typically requires the higher score, it seems likely that her getting admitted was caused in part by her retaking the test. Our evidence for this belief is the pattern that prevails in the population, but the causal claim is about what actually happened to her.

One important difference between probabilistic causal claims about populations and individuals is that the claims about individuals can only be true if the effect actually occurs. That smoking causes cancer is a true population level claim, but if Edward is a smoker who does not get cancer, it will be false that Edward's smoking caused his cancer. The population claim is true because smoking raises the chance of cancer in the population, but the claim about Edward is false, because no cancer occurred.

While some philosophers have attempted to develop probabilistic accounts of singular causal claims about individuals, others have doubted that this can be done. One reason, which we mentioned in our discussion of interpretations of probability, is that we may doubt that there really are objective single case probabilities at all. But a more serious objection in my view concerns another phenomenon which we have all experienced. Sometimes the route by which some cause leads to an effect takes an unlikely path, so an event that lowers the probability of an effect actually leads to that effect. We'll call this phenomenon, following Wesley Salmon (1980), "doing it the hard way."

Here's an example originally discussed by Deborah Rosen. I step up to the tee of a par 3 and take a swing with my golf club. I hope to strike true and land the ball on the green, but instead I slice the ball, sending it veering high and to the right. The mis-hit ball then strikes a tree beside the green, which deflects it back onto the green where, by great good fortune, it drops in the hole for a hole-in-one. My slicing the ball undeniably seems to have caused my getting the hole-in-one. Had I not sliced the ball it wouldn't have struck the tree, and had I not struck the tree it would not have deflected into the hole. At the same time, slicing the ball certainly lowers the probability of getting a hole-in-one, or even hitting the green.

It is now commonplace for sports journalism websites to have a win probability tracker. The game starts with a win probability assigned based upon the two teams' past performance, but it is updated with each play. And sometimes the swings can be quite dramatic. Suppose the Bears lead the Colts in the football game 17–14. The Bears drive down the field in the last few minutes and with one minute left find themselves on the goal line. The Bears' chance of victory has been rising with each successful play down the field, and now they seem assured to win, but as their quarterback tries to sneak across the goal line, he fumbles the football, and a Colts defensive lineman scoops it up, running the length of the field for a touchdown. Now the Colts lead 21–17 with only a few seconds left. In the space of a play, the Colts chance of victory went from 1% to 99%. The Colts really did it the hard way. On that last drive, their defense would have been trying its absolute best to stop the Bears from marching towards the goal, and they failed. But as it worked out, all of those successful plays by the Bears were part of the causal process that led to the fumble and the ultimate victory for the Colts. The general lesson here is that sometimes events that are causes may not make their effect more likely.

ASSESSING PROBABILISTIC THEORIES OF CAUSATION

Probabilistic theories of causation represent a natural extension of the regularity approaches discussed in Chapter 3. They interpret causation as difference making, specifically probability raising. Just how Humean these theories are depends upon how the probabilities are interpreted. The frequency interpretation is the most obviously Humean, since long run frequencies of outcomes are parts of the Humean mosaic. But however these probabilities are interpreted, the approach is reductive to the extent that it tries to define causes in terms of conditional probabilities.

Probabilistic theories of causation are concerned primarily with generic causal claims, but they add a new wrinkle. In the previous chapter I characterized generic causal relations as holding between types or kinds of events. We might for instance claim that heating frozen water above 0° causes it to melt. But many probabilistic causal claims, while generic, are really claims about particular populations. We wondered for instance whether a person being a woman causally influenced their rate of acceptance at Berkeley's graduate school. A causal claim about such a population doesn't always transfer to other populations. What causal influences existed at Berkeley in 1972 were likely different than what existed at Stanford in 1972, and were doubtless different than what is happening at either institution in 2022. Also, when we recognize that the individuals *within* each population can be different—applicants to Berkeley differ in their properties to a much greater degree than ice cubes do —we may be led to think that probabilistic causal claims are just ways of averaging the actual causation going on at the level of each individual.

The problem of doing it the hard way poses a fundamental difficulty for applying probability-raising theories of causation to singular causal claims. In the next two chapters we'll consider two alternatives to regularity theories that promise to provide a better analysis of singular causal claims.

SUGGESTIONS FOR FURTHER READING

Hájek (2019) is a useful introduction to philosophical interpretations of probability. Hitchcock (2021) and Williamson (2009) provide excellent overviews of probabilistic causation. Salmon (1980) also provides an overview and is one of the original sources of the "doing it the hard way" objection. Cartwright's (1979) "Causal Laws and Effective Strategies" is a classic paper that discusses Simpson's paradox and proposes a contextual unanimity criterion.

5

CAUSATION AND COUNTERFACTUALS

The theories of causation we have so far discussed have conceived of causes as things that make a difference to their effects. If there is a simple deterministic causal regularity, that difference is simply that the effect occurs whenever its cause does. More elaborate analyses like the INUS condition account allow for the relation between causes and effects to be more complicated, but still causes make a difference to their effect. Probabilistic theories are also difference-making theories, where causes make a difference by raising the probability of their effects.

But how do we understand difference-making when it applies to a single case? How, for instance, do we understand the difference-making going on when my turning the light switch causes the light to go on? The answer seems quite simple: had I not flipped the switch, the light would not have turned on. We compare the actual situation in which I flip the switch to a situation in which I didn't, and see that the flipping made a difference to the lighting. This simple idea is the germ of a counterfactual theory of causation.

Two rather different kinds of counterfactual accounts have emerged over the last fifty years. The first, which will be the focus of this chapter, originates with David Lewis (1973a). Lewis's account is a reductive analysis that defines causal dependence in terms of counterfactual dependence. The second, which we will explore in the next chapter, is

DOI: 10.4324/9781003200956-5

a non-reductive approach, known as interventionism, which explains causation and counterfactual dependence in terms of ideas about experimental interventions.

In this chapter we'll begin with a basic introduction to the logic of counterfactuals. Then we will introduce Lewis's theory, and use it to explore the strengths and weaknesses of counterfactual accounts of causation. We will also delve into a question raised by Lewis's theory, but which is important for any account of causation—namely the question of what kinds of entities can be causes and effects. So far, we have mostly assumed that causes and effects are events, but we need to say more about what events are, and also to consider whether other kinds of things, like facts, patterns, or properties, might either cause or be caused. Of particular importance for counterfactual theories are omissions. Is it possible that something that didn't happen—like my failure to lock the door—can actually cause something else to happen? We'll conclude this chapter by considering how counterfactual theories of causation are applied in the law.

COUNTERFACTUALS: THE BASICS

In Chapter 3 we discussed the logic of material conditional statements like:

(1) If I'm not at home, I'm in the office.

This statement is truth-functional, meaning its truth value is a function of the truth values of its components. Material conditionals are true just in case either the antecedent (the if part) is true or the consequent (the then part) is false. So (1) is true so long as either I am at home or I'm at the office. It would be false only if I were at neither place.

Counterfactual conditionals by contrast are not truth-functional. Suppose I walked into my office this morning and flipped on the light. Now consider these two counterfactuals:

(2) If I hadn't flipped the switch, the light wouldn't have gone on.
(3) If I hadn't flipped the switch, Elvis would have appeared.

Both these statements have antecedents and consequents that are false—that's what makes them counterfactuals—but (2) is true and (3) false. This example shows that the truth value of a counterfactual is not fixed by the truth values of its components, but given this is so, what does determine their truth value?

As a practical matter we seem able to make firm and accurate judgments about many counterfactuals. I feel confident that (2) is true and (3) is false. I feel confident that if my car had crashed into another car, there would have been noise and dented metal. I feel confident that if Hillary Clinton had won the 2016 election, that abortion would still be constitutionally protected in the United States. But the truth of many counterfactuals is more difficult to judge. Had I not gotten my current job, would I still be a philosopher? If Lincoln had not been assassinated, would we have avoided Jim Crow? Who knows?

There is clearly a close connection between causation and counterfactuals. The reason I think (2) is true and (3) is false is that I believe flipping the switch was what caused the light to go on, and I'm seriously skeptical about there being any causal connection between light switches and Elvis sightings. But the question facing philosophers is whether causes or counterfactuals are more fundamental. Do we explain what makes counterfactual statements true by appeal to causes, or do we explain what makes causal claims true by appeal to counterfactuals? Counterfactual theories of causation take the second approach.

A successful counterfactual theory of causation will require a semantics for counterfactuals, i.e., a theory that specifies truth conditions for counterfactual statements. Two similar semantics were developed in the late 1960s and early 1970s, one by Robert Stalnaker and the other by David Lewis. I'll sketch the basics of Lewis's account. Lewis (1973b) understands counterfactual claims to be about unactualized possibilities he calls **possible worlds**. Different possible worlds can be characterized by which propositions are true in them. To make this concrete, consider three propositions:

C: My flight was canceled
I: I'm in Indiana
J: I'm in Jamaica

Suppose I was flying to Jamaica but got stuck in Indiana because my flight was canceled. In this world, the actual one, both C and I are true, but J is false. The possible but non-actual world I'd hoped to be in is one where my flight wasn't canceled, and I get to Jamaica. In that world, C and I are false, but J is true.

To represent counterfactual conditionals, Lewis introduces a new symbol "□→." The counterfactual conditional *p* □→ *q* is read "If *p* were the case, then *q* would be the case." So, for instance, we write "C □→ I" to say "If my flight were canceled then I'd be in Indiana" or "~C □→ J" to say "If my flight hadn't been canceled then I'd be in Jamaica."

Because different possible worlds resemble each other to varying degrees, we can think of some worlds as being closer to each other than others. For instance, the world in which my flight is not canceled is much closer to the actual world than a world where there are no airplanes, but a world with no airplanes is still closer to the actual world than a world where humans had not evolved. Lewis uses this idea of closeness of worlds to provide truth conditions for counterfactuals:

> *p* □→ *q* is true if, and only if, *q* is true in the closest world in which *p* is true.

Let's see how this works in the case of my flight to Jamaica. We're supposing that in the actual world I'm stuck in Indiana because my flight has been canceled. But it is true that if my flight hadn't been canceled that I'd be in Jamaica (~C □→ J), because in the nearest possible world where the flight hadn't been canceled I would have made it to Jamaica. Notice that there are other possible worlds where the flight was not canceled but where I still didn't get to Jamaica. For instance, the plane might have been hijacked, or I might have been detained by the police, preventing me from boarding. But these possible worlds are further from actuality than the world where I simply board the non-canceled flight and fly to Jamaica.

Lewis spelled out this theory with considerable logical precision, and did so in a way that is consistent with commonsense judgments about which counterfactuals are true and which are false. Still, it has two features that raise suspicion. One has to do with the idea of similarity. Similarity between things in general is a matter of degrees and respects, so there seems little way to decide which thing is more similar to another. My dog Finn, an Aussiedoodle, is in some respects like Briar, my son and daughter in law's hound mix. They are of similar height and weight, and both like to run. But their fur and coloring are quite different, and while they both bark too much, they bark for different reasons and at different things. Then there's my aunt's dog Gertie, a miniature Goldendoodle. She's less than half Finn's size, but her fur is like Finn's, and her coloring is like Briar's. Dispositionally, Gertie is more like Finn, though she is much more of a lap sitter, and each has their quirks. Is Briar more like Gertie, or is Finn? There's no real answer. It depends upon what you care about. The problem is much worse for worlds than for dogs. Is the world in which I made it to Jamaica because the flight wasn't canceled more similar to this world than the world in which the flight wasn't canceled, but where I missed it because of traffic? Maybe it depends upon what caused the traffic, or upon what I would have done in Jamaica. Who knows?

Lewis understood these difficulties and did offer some general principles for judging closeness of worlds. Consider the set of possible worlds in which some event *e* that occurs in our actual world didn't occur. Lewis argued that the closest worlds to the actual will be ones that agree completely with the actual world up to the time of *e*, and diverge only right before, where through some small violation of the actual world's laws (a "small miracle"), *e* doesn't occur. Once *e* does occur, the worlds can diverge quite a lot. To see how this works, take *e* to be the assassination of Kennedy by Lee Harvey Oswald. The closest world where this event didn't happen would be a world where, by some small miracle, Oswald failed to shoot Kennedy, or missed. Call this world w^*. Perhaps in w^* the gun jammed; perhaps Oswald trembled as he pulled the trigger; perhaps there was an unexplained gust of wind. The actual world and w^* would be identical in all facts and laws up until the assassination, but would then diverge quite seriously afterward. In w^*, Kennedy would have remained president, would run for reelection, etc. There are of course other possible worlds besides w^* where Oswald could have failed to assassinate Kennedy. Oswald might not have been born, or the social circumstances that led to his radicalization might have been different, but those would have led to much earlier and broader violations of the way things were, and so would be further from the actual world than w^*. It is these judgments of closeness that help us judge the truth or falsity of counterfactuals. For instance, it seems true that if Oswald had missed, Kennedy would have run for re-election, and it seems false, if Oswald had missed, that Johnson would have become president. The latter statement would be false because, even though there are some worlds in which Kennedy survives the assassination attempt and Johnson still becomes president (say Kennedy had a heart attack), those worlds are not the closest world in which Kennedy survives.

The second difficulty with Lewis's semantics is that it is difficult to figure out what exactly possible worlds are. Lewis himself argued for a view called modal realism—the view that all these possible worlds truly exist. We happen to exist in one of these worlds, which we call the actual world, but it is no more real than the other worlds. There is a possible world very much like this world in which rather than writing in the morning I decided to take a walk. But in that world it's not really me, but is instead a counterpart, very much like me, that's taking the walk. From my counterpart's point of view, his is the actual world, and mine is a possible world in which he didn't take a walk.

Many philosophers are happy to use Lewis's semantics but to reject his metaphysics. For instance, Robert Stalnaker (1976), who is credited along with Lewis for the development of counterfactual semantics, holds that there is really only one world, the actual world, and that talk of other possible worlds, "ways the world might be," reflect certain basic modal facts about this world. What distinguishes their views is that Lewis's approach reduces modal facts about what is possible in this world to non-modal facts about what occurs or does not in nearby possible worlds, while Stalnaker takes modality as primitive. I shall not take sides on this debate here (though I would not blame the reader for giving Lewis what his opponents call the "incredulous stare"). I only want to emphasize that a counterfactual theory of causation cannot tell us what causes really are without answering the question of how we should understand possible worlds.

LEWIS'S COUNTERFACTUAL THEORY

In this section we'll introduce Lewis's (1973a) counterfactual theory of causation. It has been much discussed, criticized and revised, including by Lewis himself, but the original theory is worth exploring, as it gets at the core of counterfactual approaches to causation.

Lewis's first step is to define causal dependence between events in terms of counterfactual dependence:

> **Causal Dependence:** Event *e* causally depends on event *c* if, and only if, (a) if *c* were to occur, then *e* would occur and (b) if *c* were not to occur *e* would not occur.

Clauses (a) and (b) of this definition are counterfactual conditionals, and we can use Lewisian semantics to decide whether they are true or false.

Let's apply this definition to the light switch example: the light turning on causally depends upon flipping the switch because (a) if the switch were flipped, the light would turn on; and (b) if the switch weren't flipped, the light wouldn't turn on. Notice that causal dependence does not require that *e* and *c* are actual events. We might warn a child not to touch the pot, because it would burn them. Getting burned causally depends upon touching, but because the child was warned, neither touching nor burning occurs.

Lewis then defines causation itself in terms of chains of actual events that are causally dependent upon each other:

> **Causation** (Lewis): Event *c* causes event *e* if, and only if, *c* occurs and there exists a causal chain of events c_1, c_2, ... c_n where c_1 causally depends on *c*, c_2 causally depends on c_1, ..., and *e* causally depends upon c_n.

You may wonder why Lewis has added this extra step. Why couldn't we just say *c* causes *e* if, and only if *c* occurs and *c*'s occurrence counterfactually depends on *e*'s occurrence? The reason has to do with a problem that is the Achilles heel of counterfactual analyses of causation—**overdetermination**. Overdetermination occurs when there exist two or more potential causes, each sufficient for the effect, so that the effect does not depend causally on any of them.

Lewis's theory addresses this problem by allowing for causation without causal dependence. Here's an illustration: Suppose two assassins are sent out to assassinate the general. The primary assassin shoots the general, but the backup is there in case the primary is stopped. This kind of causal structure is called **preemption**. If the primary assassin shoots the general, then the backup hears the shot, he is preempted and does not shoot. Following Lewis's definitions, there is a chain of causally dependent events from primary assassin's firing (*p*) to the whizzing bullet (*b*) to the general's death (*e*). Because *b* preempts the secondary assassin's shot (*s*), *e* causally depends upon *b*, and because the bullet wouldn't have whizzed without primary's firing, *b* causally depends upon *p*. Per the definition of causation, *p* causes *e*, because there exists a chain of causal dependence, $p \rightarrow b \rightarrow e$. This chain exists even though *e* does not causally depend upon *p*, since *s* would have caused *e*, had *p* not preempted *s*. Lewis hoped that this causal chain technique would be enough to solve puzzles surrounding overdetermination, but the problem has proved more complicated than first imagined, and a proliferation of possible remedies have been developed (see Collins, Hall, and Paul 2004; Paul and Hall 2013).

Let's review some features and consequences of Lewis's counterfactual account. Most fundamentally, it is a theory of singular causation between events. It tells us about the conditions under which particular events that actually happened—your being born, the stock market crashing in 2008, a pipe bursting in the wall of your office—are caused by and cause other actual events. As we will discuss in the next section, events are not the only thing we typically count as causes; if so, more work will need to be done to extend the simple counterfactual account.

Lewis's account also implies that causation (though not causal dependence) is transitive, meaning that for any events, *a*, *b* and *c*, if *a* causes *b* and *b* causes *c*, then *a* causes *c*. Transitivity makes the causes of any event innumerable. Your being born depends upon you being conceived by your parents, and that demands the fertilization of a particular sperm with a particular egg. Given just how many sperm there are, just about anything about your mom's or dad's bodies, metabolism, and activities could make a difference to the creation of the genetically unique you. And their existence and meeting and mating, all at the right time and place, depends in turn about similar facts about their parents and the environment in which they lived, and so on, back beyond the beginnings of the human species, back even beyond the beginning of life itself, back to the big bang.

Some take this proliferation of causes in Lewis's theory to be problematic, but Lewis thought it was a feature rather than a bug. We have our practical reasons for identifying some things as the cause, or the principal causes of events, and we may find it useful to distinguish between causes and background conditions, but to Lewis these "principles of invidious discrimination" have no place in a theory of causation. Whether or not Lewis is right that these principles have no metaphysical import, they are essential for explaining how we reason about causes, but we will defer investigating this question until Chapter 10.

On Lewis's view, the world's causal structure is grounded in chains of causal dependencies between individual events. The truth of causal generalizations depends on these chains. Suppose events *c* and *e* are instances of event types *C* and *E*. The claim that *C*s cause *E*s can be read in lots of different ways. At its strictest, it might mean that all *C*s cause *E*s. But it also could have an implicit *ceteris paribus* clause: *C*s always cause *E*s, supposing nothing interferes. Or it could be a claim about relative frequencies, with *C*s increasing the likelihood of *E*s. These different sorts of generalizations will have different truth conditions, but all of them are made true by particular facts about the actual world's causal structure. In a reversal of regularity theories, the particular causal facts are the truthmakers for causal generalizations, rather than the existence of uniform regularities making singular causal statements true.

Given what I've said here, you might think singular causal claims don't depend upon laws. But they still do, just in an indirect way. Per Lewisian semantics, claims of causal dependence depend upon the truth of counterfactual conditionals, and the truth of these in turn depend upon whether an event makes a difference in a nearby possible world. But Lewis argues that sameness of laws is a major factor in the metric for judging similarity of worlds. A world with different laws will be quite unlike the actual world. And remember that Lewis is one of the main defenders of the Mill-Ramsey-Lewis approach to laws, which takes laws to provide the best summary of the regularities within the Humean mosaic. For this reason, Lewis's counterfactual account may be more closely allied to regularity accounts than it first appears.

THE CAUSAL RELATION AND ITS RELATA

Heretofore we've been operating on an assumption that causation is a relation between events, either particular or generic, but this assumption may incorrectly or incompletely characterize the kinds of entities that can be related as cause and effect. It's time to take a deeper dive into both the question of what kind of relation causation is, and what kinds of things may stand in this relation.

Relations are something so basic that it is difficult to offer an informative definition of what they are, but they are deeply familiar, showing up everywhere in our conceptual landscape. They express, well, "relationships" between two or more things, and these relationships can be of any kind—logical, mathematical, spatial, temporal, physical, historical, social, legal, moral.... Here are a smattering of propositions expressing relations: Paul is older than John; three is less than a million; lead is heavier than carbon; the Triassic period precedes the Jurassic period; Martha and Tom are married; Lesley is Nancy's child; Belgium is adjacent to the Netherlands.

The things that relations relate are called their **relata,** and we can characterize particular kinds of relations by the sorts of relata they have. Friendship is, for instance, a relationship that can hold among humans and arguably also among social animals. Relations like "less than" or "greater than" hold between numbers. Countries or jurisdictions can stand adjacent to each other.

Relations can also be characterized by their number of relata. My examples so far have been two-place relations, generically having the structure *Rxy*, but other relations can have three or more places. Betweenness relations (chronological, spatial, and so on) are three-place with structure *Rxyz*. For instance, in most armies, captains rank between lieutenants and majors.

Relations can also be characterized by certain logical features. For instance, some relations, like being siblings, are symmetric—if Monica is Sarah's sibling, Sarah is also Monica's sibling. Other relations, like parenthood, are asymmetric—if Kathy is Andrew's parent, Andrew cannot be Kathy's parent. Some relations are reflexive, meaning they hold between a thing and itself, while other relations are not. The ≤ relation between numbers is reflexive (e.g., $2 \leq 2$), but the < relation is not. Finally, some relations, like being an ancestor, are transitive: since Ruth is Tom's ancestor and Tom is Keith's ancestor, Ruth is also Keith's ancestor. Other relations, like being a parent are not: Ruth is Tom's parent and Tom is Keith's parent, but Ruth is not Keith's parent.

With these definitions in mind, we may say that on the most common conception of causation, the conception embodied in Lewis's counterfactual theory, causation is a two-place relation (*Cxy*) between events which is irreflexive, asymmetric, and transitive. We'll consider some objections and alternatives to this conception shortly, but before doing so, we should consider in more detail just what events are.

Intuitively an event is some doing or happening that occurs at some particular place and time. An event's temporal extent can vary widely. Some events like breaking of chemical bonds, collisions between particles, or firing of neurons last for a few nanoseconds, microseconds or milliseconds , while other events, like weddings, wars, sporting matches, are on more human scales, lasting seconds, minutes, hours, days, months or years; some events, like the formation or extinction of species or the creation of geological formations, may extend much longer. Events also vary widely in spatial extent from microscopic to galactic scales. A cell division might take place in a region of around 100 micrometers, a car crash occurs in a region of a few meters, while World War II spanned most of the globe.

Not all of the things we call causes are events in this intuitive sense. One difficulty is that we typically count among the causes of things not just things happening at a particular moment, but standing properties of things or background conditions. When we say that striking the match causes it to light, the striking of the match is clearly an event. However, the fact that oxygen is present, which doesn't seem like an event, is every bit as much required as the striking of the match for the match to light. Philosophers tend to address this problem by expanding the conception of events beyond doings and happenings to include these kinds of standing states of affairs.

An even more difficult case involves omissions. Our everyday causal explanations point to absences, failures or omissions as causes of events. The plant died because I failed to water it, or the child drowned because the bystander failed to save her. It's tempting here to think that omissions or failures could be events, but there are reasons to doubt this. For one thing, it is not clear exactly where or when these omissions or failures take place. When exactly was it that I failed to water the plants? Was it today, yesterday, or some other time? And are all of these failures the same event, or are they separate events?

These considerations have led some philosophers to suggest that events are the wrong relata for the causal relation. The most popular alternative is to take the relata to be **facts**. Facts encompass pretty much any claim about goings on in the world that we might care to make. It's a fact that certain events occurred, or that they didn't occur, or that they decreased in frequency. It's a fact that Edward is a smoker, or that gas prices went up, or that I forgot to turn off the light, or that Elvis wasn't in the building.

But facts also have some strange properties that make them less than ideal candidates for relata. Facts, unlike events, have no location in space and time. The fact that I was born in November is a fact in November, but also in March, and it will continue to be so forever. It is also not located where I was born; It's true everywhere, or, as philosophers like to say, transcendent. The supposition that causation is a relation between facts thus runs strongly counter to the intuition that causes act at particular times and places.

Let's return to events and see some of the difficulties that can arise when we use them in the counterfactual theory. Consider the following scenario. Your office has planned the big holiday party and the pandemic hits, locking everybody out of the office. Your party animal colleague Antonio hits on the idea that instead of meeting at the planned venue, everyone should stay at home, and at the scheduled time, grab a cocktail and get online for internet party games. Everyone joins and has a great time. Question: Did Covid cause the party to be canceled? The answer hangs on whether you think the online party was the same party as the party that would have happened at the office.

Intuitively it seems that some features of events are essential, while others are accidental. If no one shows up, the party doesn't occur, but if one guest is late, or if you burn the appetizers, that hardly stops the party. The line between what is essential or not is vague at best, and seems likely not to reflect any objective facts about the world, but rather some kind of individual preference or social agreement about what we take to be essential.

Here's another example. Suppose your dad is in the final stages of cancer and is in hospice care. He is in great pain and not coherent, so the nurse, with your consent and following your dad's earlier directives, administers morphine. He lapses into sleep and in a few hours passes peacefully, 12 hours earlier than he would have, had the morphine not been administered. Question: Did the morphine (and by extension the nurse, or even you) cause your father's death? Certainly the death was hastened, but is it the same death as the one he would have died a few hours or days later? The law doesn't consider palliative administration of drugs at end of life to be murder, but perhaps strictly speaking it is. And how we judge this depends crucially upon the context. Were your dad likely to live for a month more, and especially if your Dad had not given consent, we would consider the nurse to have killed him. Proximity in time matters, but that is not all of it. Suppose Thug attacks and kills Victim, who is asleep in their home. A couple of hours later, Victim's house bursts into flames because of a tragic but accidental gas explosion. Victim, we'll suppose, would have died in that explosion. But unlike in the hospice case, we would clearly hold that Thug caused, and not just hastened, Victim's death. Victim's death is a different event than the death he would have had had Thug not killed him.

Considering these kinds of examples tempts many philosophers to adopt a fine-grained view of events. On fine-grained views, lots of features are essential to the event, so that it doesn't take much to stop an event from happening. On a fine-grained view, the party was in fact canceled, and replaced by a new party. On the fine-grained view, it was really the morphine that caused your dad's death, just as it was Thug's attack that caused Victim's death. The differences in how we treat these things do not reflect causal differences, but instead practical, legal or moral differences. Tempting though this is, the strategy of going fine grained is not without its problems. Consider again the party. You might think the actual causal facts are clear: these six people showed up, this is what they ate and drank, this what they did with each other, and call it what you may. But even this fine-grained description could get much finer. At what precise time did they show up? What neurons were firing in their brains, etc.? Going down the path of fragility leaves us without any genuine events beyond some mythic microphysical events, and leaves us far away from any practical understanding of causation.

Finally, here is a somewhat different kind of problem. Suppose my snoring causes my wife to wake, but suppose also that had I turned in bed slightly, I would have still snored, but snored more softly, and that my quiet snoring would not have waked her. Did my snoring then really cause my wife to wake? There are two ways we might interpret this case. One way would be to treat my loud snoring and my softer snoring as distinct events. The actual loud snoring caused my wife's waking, because had it not happened, i.e., had I either not snored at all or snored more softly, she wouldn't have woke. Alternatively, we might say that it is not the event (the snoring) that caused it, but a feature of that event (its loudness). This, however, would be to reject the view that causation is a relation between events, and instead to say that it is a relation of aspects of events (Paul 2004).

Another possible kind of relatum for causal relations are quantities or magnitudes. By focusing on events, philosophers have focused on yes-or-no causal questions. Did event *c* occur or not? Does *e* causally depend upon *c*, or not? But many causal claims express quantitative dependencies of various kinds. As a very simple example, consider the relationship between a car's steering wheel and its front tires. We can describe this in event-talk—turning the wheel turns the tires—but it is more accurate and informative to express the causal relationship in terms of the quantitative relation between the angle of the wheel and the angle of the tires.

Many physical laws and causal generalizations in the sciences are expressed as quantitative relations between magnitudes. Take the ideal gas law, $PV=nRT$. Although equations like this don't in themselves have the kind of directionality that we expect between cause and effect, they can be read causally in various ways. For instance, if we hold volume V fixed at v_0, then changing the temperature causes a change in the pressure, $\Delta T \rightarrow (v_0/nR)\ \Delta P$. Other examples of quantitative causal relationships are relationships between environmental variables like temperature or sunlight and growth rates in plants, or relationships between dosages of drugs and clinical effects. Some of these relationships may hold between individual quantities (say between the temperature and pressure of a gas within a single experiment), but others may be population level claims. For instance, we might calculate the average effect on crop yield of applying various amounts of fertilizer. There can also be quantitative relationships between a quantitative cause and the probability of effects. For instance, the probability of cancer may increase with the amount of exposure to carcinogens. In our earlier discussion of whether smoking causes cancer, I've acted as if smoking were a yes-no variable, but in fact the quantity and timing of smoking (e.g., amount smoked per day and number of years of smoking) has a significant impact on the probability of having cancer, its severity, and the timing of its onset.

Up until now I have assumed that causation is a two-place relation between a cause and an effect. But as I mentioned above, relations need not be limited to two places, and some philosophers have proposed that causation is actually a three or four place relation. To see why they might think so, consider another causal story.

Suppose Charlie stole a bike and then crashes into another bicycle. Did his stealing the bike cause the accident? That depends. On the one hand, if he didn't have the bike, he wouldn't have crashed; but on the other, if he had bought the bike he still would (we'll suppose) have crashed. The question seems to hang on an implicit contrast. Are we asking whether the crash was caused by his stealing *the bike,* or are we asking whether it was caused by his *stealing* the bike? (Of course, it might be that his *stealing* does matter, because it might, for instance, have accounted for him rushing through an intersection.)

One way to respond to this puzzle is, following Paul's suggestion, to retreat from the idea that causation is a relation between events, for what seems to be causing the accident is a certain aspect of the theft, that it was a bike that was stolen, rather than the event as such. Another way, proposed by Jonathan Schaffer (2005) is to treat causation as a four-place relation: rather than it simply being the case that c causes e, it turns out that c rather than c^*, causes e rather than e^*. In our example, for instance, (a) stealing the bike rather than stealing something else is causally related to Charlie having an accident rather than walking home safely; but (b) stealing the bike rather than buying it is not causally related to Charlie having an accident rather than walking home safely.

I'll close this discussion by sharing a proposal I've made for what I call a canonical form for singular causal claims (Glennan 2017). The general form is this:

> *c* causes *e* in virtue of relevant feature *p*.

The idea behind this form is to preserve the idea that causation is a relation between events, while recognizing that this relation only obtains because some aspect of those events is causally relevant to the outcome. To use a simple example, we can say that turning the key (*c*) causes the lock to open (*e*) in virtue of the shape of the key (*p*). This formulation allows us to single out some aspects of the event (the key's shape) as relevant, while leaving out irrelevant aspects like the key's color. Or, you could apply this to the case of Charlie stealing the bike: Charlie's theft caused the accident in virtue of the fact that the item stolen was a bike.

You could read my proposal as claiming that causation is a three-place relation between two events and a feature (or aspect), though I prefer to avoid that interpretation. Personally I have doubts about whether there is a unique relation called causation, so I'm not going to spend too much time trying to identify its logical structure. Instead, I think it's best simply to notice that there are lots of different kinds of causal claims, and recognize that different ones of these claims will have different facts that make them true. We cannot decide these matters here. It's enough for the purposes of this book to be aware that causation does not always appear to be a straightforward two-place relation between events.

COUNTERFACTUALS AND THE LAW

One of the most important applications of counterfactual approaches to causation is found in legal reasoning. Having standards for establishing causal responsibility is essential for both civil and criminal law, because one cannot hold someone to account for a harm that they did not cause. Within the Anglo-American legal tradition, establishing causal responsibility is a two-step process. First one must show that a certain action (or omission) is a **cause in fact** of the harm, and second, one must show that the cause in fact is also a **proximate** or **legal** cause.

To be a cause in fact, an action must be necessary for the harm to occur. This necessity is established using the **but-for test**, which in reality is nothing more than the principle that the harm must counterfactually depend on the action. For example, we say that a driver (call her Alice) running a red light is a cause in fact of the accident, because, but for her running the red light, the accident would not have occurred. There are cases where something the law counts as a cause in fact fails the but-for test, and these mirror problem cases for counterfactual accounts of causation, like overdetermination and preemption. If for instance, two gangsters both fired shots at a victim, each of which was sufficient to kill the victim, then neither individual shot was necessary. Still, each of the shots are treated by the law as actual causes.

The reason one needs the second step of identifying a proximate cause is that the but-for test will identify many causes in fact that are not relevant to the law. For instance, if the accident took place as Alice was returning home from a party to which she had been invited, it is plausible that, but for the invitation, she wouldn't have struck the car. It is also plausible that, but for the host serving alcohol at the party, Alice wouldn't have gotten drunk and then she wouldn't have run the red light. Additionally, the accident depended not just on Alice's running the red light, but also on the other driver (let's call him Bob) driving through the intersection at just the right time. But for Bob deciding to go to the drug store to pick up his prescription, there would have been no accident.

The concept of proximate cause is meant to select one or more of these many causes in fact for which a party can be held civilly or criminally liable. There is no simple characterization of what makes a cause proximate, but rather a set of tests that lawyers can argue about as they seek to fix liability. They include the principle that a cause's connection to the harm must be direct and that its effect must be foreseeable. Using these tests, it seems pretty clear that Alice's running of the red light is the proximate cause of the accident, while Bob's driving through the intersection was not. Bob could not foresee his driving through the green light would cause an accident while Alice could (or at least should) foresee that running a red light would. Similarly one cannot blame the accident on Bob for deciding to go to the drug store or on Alice's friend inviting her to the party, since these causes in fact were both more distant from the accident, and the effects of these acts on the accident could not be foreseen.

One respect in which the legal concept of cause in fact differs from Lewis's conception of cause is that Lewis takes causes to be events, while the law takes proximate causes to be either actions or omissions.

The relation between actions and events is a matter of dispute, which we shall consider in more detail in Chapter 8, but for the moment we can think of actions as a special kind of event which involves a choice by a human agent. Driving through a red light is an action by this criterion, since it involves an agent, while a tree branch breaking in the wind is an event, but not an action. The reason for the focus on actions is obvious; it is ultimately only people, or people-like entities like corporations, who can take actions and be held to account. We can blame Alice for running the red light, but we can't blame the tree for its branch falling.

Omissions are not events, and therefore do not count as causes according to Lewis's theory, but they are essential to the legal concept of cause, since often we want to hold agents to account not just for their actions, but for their failures to act. The law classifies such failures as negligence. For instance, a parent can be held criminally liable for their failure to lock a gun cabinet if their child shoots someone with a gun found in the cabinet. Their failure to lock the cabinet is interpreted both as a cause in fact, meeting the but-for test, and also as a proximate cause, because the connection to the harm was direct and foreseeable.

Causation in the law is a place where we see clearly the interplay of fact and convention in determining what counts as a cause. It's clear that conventional elements enter into the selection of proximate causes. To decide whether an action was negligent requires a judgment that the defendant could reasonably foresee the risk of harm from the action, and what might count as reasonable depends on standards of reasonableness that cannot be read off the world's causal structure. And there are some standards where the conventionality is apparent on its face. For instance, it is still the case in many jurisdictions that an action cannot be a cause in fact of a death that occurred more than a year and a day after the action occurred. This principle has practical justifications, but the line is arbitrary, and can and has been changed.

Conventional elements also enter into the determination of causes in fact. Consider again the case of the nurse administering morphine. If we judge the death your father had after administration of morphine to be the same death as he would have had the next morning without the morphine, then the nurse did not in fact cause the death, but only hastened it and made it less painful. If instead we count it as a different death, then he caused it. Clearly such judgments rely upon policies and conventions, more than simply matters of fact.

SUGGESTIONS FOR FURTHER READING

David Lewis's (1973) "Causation" is the foundational work for contemporary counterfactual theories of causation, and while it is sometimes technical, it is worth the struggle. The collection of essays, *Causation and Counterfactuals* (Collins, Hall, and Paul 2004), includes important papers developing counterfactual theories, including a significantly revised theory from Lewis himself. Paul and Hall's *Causation: A User's Guide* (2013) is a synoptic introduction to contemporary counterfactual theories, with an emphasis on metaphysical questions.

Ehring's "Causal Relata" (2009) provides an excellent introduction to events and their alternatives as candidates for causal relata. Moore's "Causation in the Law" (2019) provides a succinct overview of the philosophical and legal theory that undergirds the distinction between causes in fact and proximate causes, as well as other issues about causation that arise in legal contexts.

6

CAUSAL MODELS AND INTERVENTIONS

In the last chapter we discussed David Lewis's counterfactual theory of causation, an approach chiefly aimed at providing an account of commonsense judgments about singular causal relations. In this chapter we will look at a second kind of counterfactual theory, motivated more by scientific considerations, and concerned primarily with understanding generic causal relationships. This approach, especially associated with the work of computer scientist Judea Pearl (2009) and philosopher of science James Woodward (2003), is a nonreductive approach that draws its inspiration from the role of experiment in acquiring causal knowledge. This approach, often called interventionism, understands causes as things that can, at least in principle, be manipulated by interventions to make things happen in the world.

The interventionist approach to causation is closely tied to a set of mathematical techniques for representing the world's causal structure that go by the name of "causal modeling." Causal modeling has developed rapidly over the last thirty years and has been central to rehabilitating the reputation of causal reasoning in the sciences. We'll need therefore to begin this chapter with a non-technical look into how causal modeling works.

CAUSAL MODELS

The theories we have so far discussed have tried to identify what is required for x to be a cause of y, what the nature of the connection is

DOI: 10.4324/9781003200956-6

between x and y, and what kinds of things x and y can be. But implicit in that investigation has been something that deserves more explicit attention—the fact that the world has **causal structure**. There is much more to that structure than the mere fact that countless pairs of events or facts are related as cause and effect. We understand that causes and effects form a vast network, that between causes and effects there may be long chains of intermediate causes, that causes will have both immediate and longer term effects, that some causes act in concert and others act against each other, that effects generally require the joint operations of many causes, and so on. Many of the questions we seek to understand about causes go beyond the mere question of whether something is a cause of something else, to explore more complicated questions about how large the effect is, how various causes of the effect interact, and so on. To answer these questions we need **causal models**, which are, broadly speaking, ways to represent aspects of the world's causal structure.

The causal modeling techniques described in this section are a relatively recent development that has been the subject of interdisciplinary research in statistics, computer science and philosophy over the last thirty years or so. This research has led to a set of mathematical techniques for constructing causal models and testing them against empirical data. I shall introduce some basic ideas about these causal models here, and will return to discuss their connection to causal inference in Chapter 9.

While these new causal modeling approaches have proved very fruitful, they are but one of a much larger family of techniques that scientists use to represent aspects of the world's causal structure. Whether explicitly or not, the vast majority of models used by natural and social scientists, engineers, and policy makers to describe systems and processes in their domain are in this broader sense causal models.

The important thing to keep in mind in evaluating any model is that it is a **representation**. Like any representation—whether a painting, a photograph, a map, a recording, or a history—it includes some things about what it represents, simplifies or distorts others, and leaves many things out. This fact about models implies that there is no perfect or complete model of the world's causal structure, or really of any part of it. Different models will be useful for different purposes. As we look at models of causal structure, it's important to be on the lookout for what is left out, and to consider whether any of these things should be part of the causal story.

The two diagrams in Figure 2 are called **directed acyclic graphs** (DAGs). The capital letters are variables, and the arrows represent causal connections between these variables. Figure 2a pictures a common cause structure. We might for instance take *P* to represent atmospheric pressure, *B* to represent a barometer reading, and *S* to represent a storm. Because the letters are variables, they can take different values. The values can either be continuously varying magnitudes, like the atmospheric pressure, or be binary variables indicating the occurrence or non-occurrence of events like a storm.

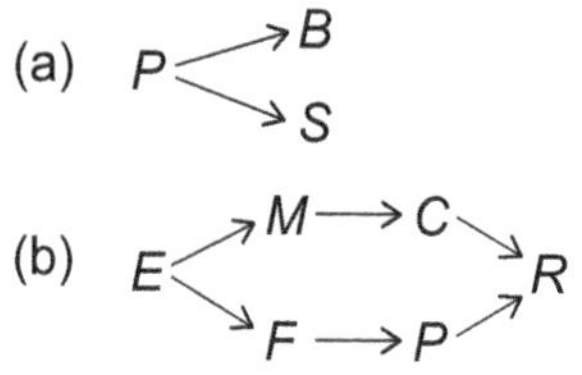

FIGURE 2. DAG models of two causal structures

Figure 2b is a slightly more complicated causal structure. We could use it to represent a scenario I'll call "fake news": Suppose candidate Garcia is elected in a close race (*E*). The results are reported in the media (*M*), leading to celebrations (*C*) by Garcia's supporters. At the same time, opposition media spreads false allegations of election fraud (*F*) which lead to protests by Garcia's opponents. The combination of protesters and celebrators in the street causes a riot (*R*).

These graphs are directed because their arrows point only one way, and acyclic, because paths traced in the direction of the arrows don't loop backward. The diagrams are read from left to right, with the earlier events causally influencing later events. It can be helpful to use genealogical terms, like ancestor, descendant, parent, child and sibling to describe relationships between various nodes in a graph. For instance, in Figure 2b, *E* is the parent of *M* and *F* (who are siblings), and is the ancestor of all other nodes.

DAGs can represent several important features of causal structure. They allow single causes to have multiple effects, and single effects to depend upon multiple causes. They also allow us to distinguish between direct effects represented by arrows, and indirect effects, mediated by intervening variables. What DAGs don't yet tell you is how exactly the causes on the left end of arrows determine the effects on the right. The standard way to provide this information is with **structural equations**.

Structural equations provide quantitative specifications of the relations between variables connected by arrows. The following set of structural equations is a complete specification of the fake news scenario:

$E = 1$
$M = E$
$F = E$
$C = R$
$P = F$
$R = C \,\&\, P$

The variables in this model are binary, with the value 1 representing the occurrence of an event and 0 representing its non-occurrence. Because these equations are used to represent causal dependencies, the equations must be read functionally, with the effect variable on the left being dependent upon the cause variables on the right. So, for instance, in the second equation *E* (Garcia winning the election) determines *M* (the media reports), and not the other way around.

E is called an exogenous variable because its value is not determined by the model. The other values are determined via the equations, from *E*. In this case the relationship is quite simple, with each event triggering the next, except that, for *R* to occur, both *C* and *P* must occur. This set of equations actually contains all of the information encoded in the diagram, so the diagram becomes strictly speaking unnecessary. You can construct the diagram from the equations by putting arrows from any variables on the right side of an equation into the dependent variable on its left side.

One of the most useful features of causal models is that they encode information about counterfactuals. We can use the DAG to calculate the consequent of a counterfactual, by simply setting a variable in the model to its counterfactual value, and then seeing if the consequent is true. We'll use a notation of Judea Pearl's called the Do operator. Consider the statement, "If Garcia had lost the election, there would have been no riot." To evaluate this, we just set *E* to 0, Do(*E*=0), and use the equations to calculate the value of *R*, which then come out as 0. This tells us that if Garcia had lost the election, there would have been no riot. Similarly, we can consider what would have happened if there had been no fraud allegations by calculating the value of *R*, given Do(*F*=0). Again, the model tells us that there would have been no riot.

If one has a causal model, this procedure gives a semantics for counterfactuals that is much more straightforward than Lewis's similarity-based semantics. There's no need to evaluate a counterfactual by considering the truth of claims in nearby possible worlds. One simply uses the Do operator to make the antecedent of the counterfactual true, and then reads the result off the model. This approach is at the heart of the second counterfactual theory of causation, interventionism, which we will discuss in the last part of this chapter.

Structural equation models (and DAGs) are not limited to binary variables, and can use quantitative variables to represent causal relations between magnitudes. As an example, consider a model of a furnace attached to a thermostat. For the furnace to work (*F*), there must be power coming to the house (*P*) and the breaker must be turned on (*B*). If the Furnace is working, the inside temperature (*I*) will be the value set by the thermostat *T*. If it isn't working, the inside temperature will be equal to the exterior temperature (*E*). This structure is captured in two equations:

$$F = B \,\&\, P$$
$$I = FT + (1\text{-}F)E$$

The second equation uses a mathematical trick to represent a conditional statement; if F=1, then I=T; otherwise I=E. The resulting graph is given in Figure 3.

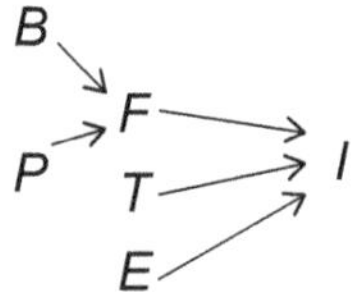

FIGURE 3. A causal model of a furnace

Causal models are primarily representations of generic or type-level causal relationships. The furnace model tells you generally how interior temperature depends upon the exterior temperature, the thermostat settings and the other variables. It is only by setting the exogenous variables to values representing an actual situation on some occasion that you can represent singular causal relationships. For instance, if on some occasion the thermostat is set to 65°, the outside temperature is 40°, and the breaker and the power are on, then the exogenous variables (B=1, P=1, T=65°, E=40°), determine that

F=1 and *I*=65°, i.e., that is the furnace is on and the interior temperature is 65°. Further, if the external temperature drops to 20°, the furnace will still keep the interior temperature at 65°, but if the power goes out, the interior temperature will drop to whatever the actual exterior temperature is.

The furnace model is a greatly oversimplified representation of the causal structure of any actual heating system. According to this model, the furnace also can cool air (maybe it's a heat pump), since if you set *T* to less than *E*, *I* will still equal *T*. Also, the model doesn't represent the inevitable time lags between changes in variables, as systems come to thermodynamic equilibrium. If I were to lose power tonight, my house is well enough insulated that it would take many hours to reach the colder outside temperature. Finally, you may notice something odd about the model's depiction of the relation between the thermostat and the furnace. The model shows the thermostat setting and the furnace having power as two causes of the internal temperature coming to some value—and surely that is true—but what it does not show is that the way the thermostat effects the internal temperature is by switching the furnace on and off.

Causal models can be used to represent both very specific bits of the world's causal structure, or to represent a class of causal structures that occur in different times and places. For instance, the furnace model in Figure 3 is a causal model of my house's heating system, but it also presumably provides a reasonable model of a great many other houses' heating systems. In using causal models, there is typically a tradeoff between the specificity of a model and its scope. There are doubtless, for instance, many detailed features of my house's heating and cooling system that make it different from the one in my neighbor's house. A coarse-grained model might work equally well for both of our houses, and will tell us some essentials about the effect of interventions—enough to figure out how to set the temperature; but for other purposes—say, to undertake repairs—more specific models will be needed.

The structural equation models we've considered so far have been deterministic, but the most common applications of these models use probabilities. Figure 4 is a DAG representing probabilistic causal relationships between smoking (*S*), heart disease (*H*), lung cancer (*L*), and death (*D*). The variables in this example are **random variables** representing the distribution of values for these variables within some population. Associated with each link, we can write a structural equation that includes an error term *U*. For instance:

$$D=f(H, L) + U$$

expresses the direct causal influence of both heart disease and lung cancer on death. The error term U represents everything left out of the model that has an influence on deaths, as well as any measurement error. Hidden in the error term are other causes of death (strokes, drownings, car accidents...) that may or may not be causally connected to smoking, as well as other unmodeled causal factors that explain why some who have lung cancer or heart disease survive, while others die.

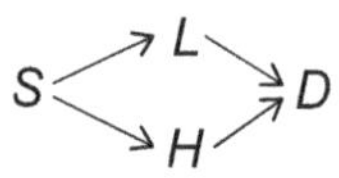

FIGURE 4. A causal model of the effects of smoking

This model is obviously fragmentary, but it is enough to say something about how smoking causes death—namely that it is mediated by at least two kinds of diseases—heart disease and lung cancer. Such a model is not justified simply by correlations between smoking and these diseases. Such correlations could potentially be explained away by common causes, and without adding additional variables, most of the causal structure responsible for these diseases will be left unexplained.

OTHER REPRESENTATIONS OF CAUSAL STRUCTURE

Useful as DAGs are, they cannot represent certain aspects of causal structure. One important case is feedback loops. In general, feedback loops occur when a change in some variable A causes a change in a variable B, and then that change in B, either directly, or indirectly via a causal chain causes a further change in A. The two simplest kinds of feedback, illustrated in Figure 5 are positive feedback and negative feedback. A simple example of a negative feedback loop is the loop between a thermostat and a furnace. A thermostat has a temperature sensor connected to a switch. When the ambient temperature falls some increment below the desired temperature, say 65°, the thermostat switches on the furnace. Heat from the furnace raises the ambient temperature until the thermostat detects that the desired temperature has been reached, at which point the thermostat switches the furnace off.

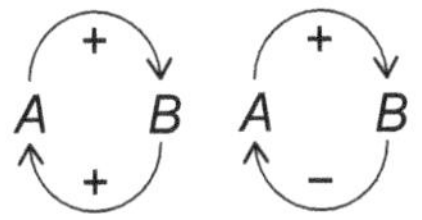

FIGURE 5. Positive and negative feedback loops

While negative feedback loops help maintain systems at equilibrium, positive feedback loops are causal systems where the feedback accelerates change. A simple example of positive feedback is a bank run. In a bank run, some set of depositors begin withdrawing funds in response to some concern over bank solvency. This pattern of withdrawal is detected by other depositors, creating increased concerns about solvency, which leads to increased withdrawals, and so on, until such time as the bank becomes actually insolvent:

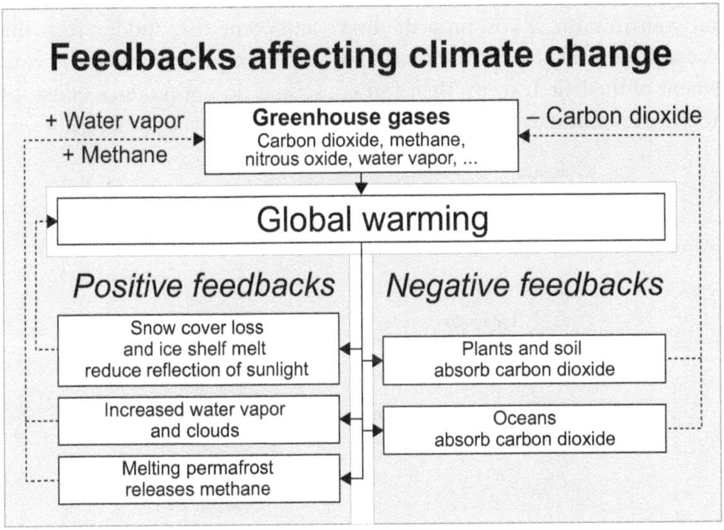

FIGURE 6. Feedback loops in climate change (By RCraig09: Own work, CC BY-SA 4.0, https://commons.wikimedia.org/w/index.php?curid=121130309)

Figure 6 illustrates a more complex system of feedback that accounts for the increased speed of global warming. Global warming induces a variety of environmental and ecological changes. Some of these, like release of methane from melting permafrost, increase greenhouse gases and thereby accelerate global warming, while others, like potential increases in carbon dioxide-absorbing plant life, may dampen it. The scientific consensus on global warming is that positive feedback will exceed negative feedback, so that even with zero greenhouse emissions, global temperature will continue to rise.

While cyclic graphs are helpful in representing the causal structure of systems with feedback loops, they don't imply anything weird about the direction of causation. A backward arrow in a feedback loop doesn't imply going backward in time. In feedback loops a variable A at time t_0 influences a variable B at time t_1 which then influences A at t_2, and so on.

The examples we've so far considered of causal structure and its models have focused on complex processes of action or influence. But sometimes the best way to characterize the world's causal structure is not by modeling these processes, but by describing what constrains them. Here is an example of a constraint: when you buckle on your seat belt, it places a constraint on your motion. If you press the brake and come to a sudden stop, this causes your body to move forward, but you are kept in place by the constraint of the belt. It seems then that constraints do not produce effects by themselves, but they limit, prevent, or direct the effects of other causes.

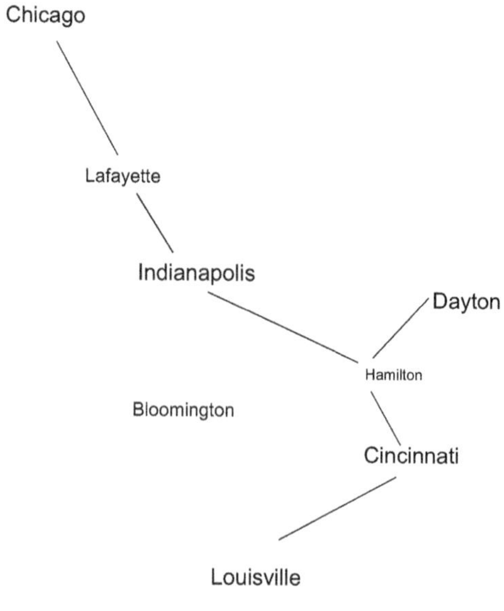

FIGURE 7. A freight railway map

Many representations of causal structure are in fact representations of constraints. Figure 7 represents a fragment of the freight rail network around my hometown of Indianapolis, which shows you the paths you must follow if you want to ship freight. For instance, if you want to ship freight from either Dayton or Cincinnati to Chicago, it must go through a little railroad junction named Hamilton, and thence through Indianapolis and Lafayette. You'll also notice that to get from Louisville to Indianapolis, there's no direct route, but you must track back east to Cincinnati and go from there. I've added Bloomington to the map, but it is isolated, unreachable nowadays by rail.

Once you start looking for them, constraints are everywhere. These constraints are imposed by train tracks and station locations. But there are also, among other things, biological, social, political, economic, legal, and moral constraints, all of which may shape the direction and ends of causal processes. Philosophers disagree about whether constraints should be thought of as kinds of causes ("structural causes") or if they should be thought of as non-causal. However this matter is resolved, it is clear that constraints play an important role in explaining why and how things do or don't happen.

CAUSATION AND INTERVENTION

One of the most valuable things about causal models is that they provide information about what would happen if someone were to make changes to variables represented in the model. The furnace model in Figure 3, for instance, tells you that were you to turn off the circuit breaker (and thence the furnace), the inside temperature would become the same as the outside temperature. Of course, these counterfactual claims would only hold true to the extent that the model adequately captures the relevant portion of the world's causal structure. If for instance there were an unmodeled second furnace system, or if there was a backup generator, turning off the circuit breaker wouldn't have the predicted effect. But if the model is adequate, it gives you a simple way of evaluating the truth of counterfactuals about interventions into a system. This fact suggests that a counterfactual theory of causation could be grounded in the idea of intervention (Pearl 2009; Woodward 2003).

Interventionism begins from the commonplace observation that we use causal knowledge to make things happen in the world. Causes to the interventionist are the knobs we can turn to bring about certain effects. If I turn the dial on the stove, the flame gets bigger; if I flip the switch, the lamp goes on; If I water the lawn, it grows; and if I fail to water it, it dies. These kinds of interventions are what the Do operator represents.

We come to understand causation through intervention as we learn how to manipulate things (toys, food, people…) to accomplish our goals, and experimental methods are just refined versions of these childhood principles. The basic structure of controlled experiments involves interventions; one divides a population into experimental and control groups to test for a causal connection between two properties *X* and *Y* of members of that population. In the experimental group the experimenter wiggles (intervenes on) the value of *X* and checks the value of *Y*. They then compare it to the control group where *X* wasn't wiggled, to see if the wiggling made a difference to *Y*. If it did, the experimenter judges *X* a cause of *Y*. We'll get to some complications in a moment, but the gist of the interventionist theory is that what it means to say that *X* causes *Y* is simply that if you wiggle *X*, you will thereby wiggle *Y*.

An interventionist theory is a counterfactual theory of causation, because it grounds causal claims in facts about *possible* interventions. If I assert, for instance, that the oysters I ate caused my indigestion, I mean that had I intervened on my dinner by changing from oysters to something else, then I wouldn't have had the indigestion. Obviously I can't go back and change what I actually did, but if oysters caused my indigestion, I know that another dining choice would have made a difference.

But just what is an intervention? A natural starting place connects intervention with human action. In the example above, the intervention would simply be the act of choosing a different meal. Similarly, if the gas leak caused the explosion, this means that if we had acted to fix the gas leak, the explosion wouldn't have occurred. But tying interventions to human action seems too anthropocentric. Causes and effects would exist in the absence of humans, and certainly many things we count as causes and effects are beyond the reach of human action. An alternative approach, developed chiefly by Judea Pearl and Jim Woodward, seeks to define interventions as variables within causal models that meet certain conditions. The key idea is that if *X* really is a cause of *Y*, it should be possible to change *Y* by a surgical intervention that changes *X* and nothing else.

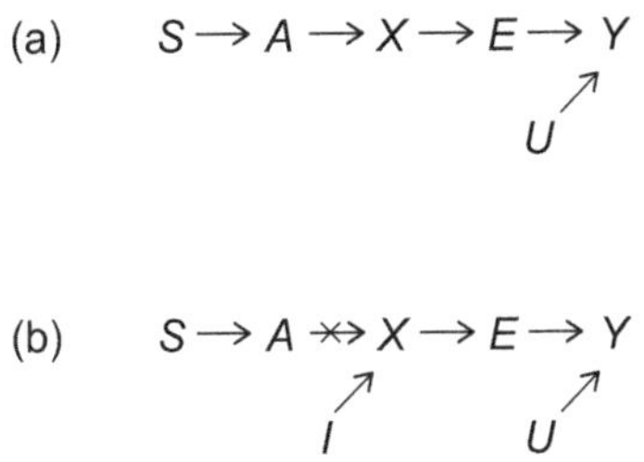

FIGURE 8. Causal model of a flying plane

To explain this more precisely, let's apply it to a model of a flying plane, shown in Figure 8. In the absence of interventions (Figure 8a), a set of sensors *S* cause the autopilot to move the stick *X* into a position that sets the elevators at a certain position *E*. *E*, in combination with a set of other factors (like airspeed, thrust, and so on—all modeled by the variable *U*—determine the angle of flight *Y*. Figure 8b shows the effect of a pilot's intervention *I*, which overrides the autopilot and manually moves the stick, thereby changing the elevator position and angle of flight.

For the pilot's intervention *I* on *X* to be *ideal* with respect to *Y*, the following conditions must hold.

1) *I* must function as a switch, so that *X* depends only on *I*, breaking dependencies on other variables. In this example, grabbing the stick overrides the autopilot, indicated in the diagram by the "x" marked through the arrow between autopilot and the stick.
2) *I* must influence *Y* only through the path of changing *X*. *I* cannot somehow interfere with intermediates between *X* and *Y* (like *E*); nor can it change anything else U upon which *Y* might depend. Nor can it change *Y* directly. Notice that there are no arrows connecting *I* to *E*, *U*, or *Y*.

Often interventions aren't ideal. In this case, imagine that the copilot, seeing the pilot pull on the stick, takes some other actions (increasing power to the engine, engaging some backup system, etc.) that also influences *Y*. If this happened, the pilot's pulling the stick would not be ideal, because the intervention would affect *Y* through more than one route (both through *X* and *U*).

Ideal interventions are analogous to the kinds of interventions performed in well-designed experiments. In a controlled experiment, care is taken to assure that control and experimental groups differ in treatment only with respect to one variable. The point, for instance, of giving placebos and blinding is to make sure that the only variable changed is the pharmacological agent. At the same time this account of interventions avoids the charge of anthropocentrism, since interventions can meet (or approximate) these conditions without being human actions). For instance, a cold snap can count as an "intervention" which freezes a crop, or a stroke as an "intervention" which blocks blood supply to some part of the brain.

ASSESSING INTERVENTIONISM

The interventionist account of causation can be interpreted either epistemically or semantically. In other words, it can be seen as an idealized account of the conditions under which we can infer that *X* causes *Y* (epistemic) or it can be understood as an account of what it means to say that *X* causes *Y* (semantic). In the diving plane scenario, the natural reading is epistemic. If the position of the stick is causing the plane to dive, the pilot's intervening to pull on the stick will change the plane's path. You can imagine a pilot actually doing this as she tries to diagnose the cause of the dive. But we can also read this semantically. What it means to say that *X* causes *Y* is that if there were an intervention on *X*, it would change *Y*. It is not an accident that interventionist counterfactuals have these two readings. Advocates of the theory see it as a virtue of the theory that there is a tight connection between what we think causal and counterfactual claims mean, and how we come to believe they are true.

One thing that the interventionist account clearly *is not* is a reductive analysis of causation. It instead gives an account of what causes are generally by reference to a specific and well understood kind of causing—an intervention. What does it mean to say that importation of Japanese Chestnut trees caused the death of billions of American Chestnuts? It means just that these trees would not have died (or at least many fewer of them would have died) had someone intervened to block the importation of Japanese Chestnuts. This answer is helpful to the extent that it shows how our idea of cause is connected to actual and possible interventions, but intervention of course is a causal concept.

The interventionist account thus stands in stark contrast to Lewis's counterfactual theory of causation. Lewis's reductive analysis gave a set of truth conditions for counterfactual and causal claims that was grounded in the truth of non-causal claims about other possible worlds. Pearl and Woodward though are unapologetic about the nonreductive approach. The Lewis account appeals to a similarity relation across possible worlds which is vague at best, and it may be the case that what makes worlds similar or not is their *causal* structure, in which case causation sneaks in again, and the reduction fails. Advocates of the interventionist approach believe they can draw a straight line between causation and experiment, and that the causal notions they assume are decidedly less obscure than the apparatus Lewis uses to explain away those notions.

The interventionist approach to causation has proved very influential in the last twenty years among philosophers of science and scientists interested in causal inference and explanation, largely because it comports so well with the way that natural and social scientists search for and identify causes. Nonetheless, the theory leaves important questions unanswered. It tells us that *X* causes *Y* if we can intervene on *X* to change *Y*. But this fact seems more to be evidence of a causal connection rather than constitutive of causation itself. We may establish the causal connection between a switch and the light by experimenting with the switch, but we assume the reason that the switch can turn on the light is because there is some mechanism or process that connects the switch to the light. In the next chapter we will turn to a collection of approaches to causation that try to spell out what such processes or mechanisms might be. These approaches reject Hume's skepticism about the idea of necessary connection, and aim to offer a scientifically informed account of how causes are connected to their effects.

SUGGESTIONS FOR FURTHER READING

A good starting point for further information on causal models is Hitchcock's (2023) article in the *Stanford Encyclopedia of Philosophy*. Pearl's classic *Causality (2009)* has technical parts, but contains much that is informative and accessible. Woodward's *Making Things Happen* (2003) is the standard account of the interventionist approach.

7

CAUSAL PROCESSES AND MECHANISMS

In Chapter 1 I suggested that theories of causation can be classified into two categories, those that conceive of causes as things that make a difference to their effects, and those that conceive of causes as being connected to their effects. The theories we have considered in the last four chapters all fit within the difference-making category, since all of them in some way define some kind of comparison of situations where the putative cause occurs and doesn't occur. For classic regularity theories, the difference is expressed in the fact that a cause is necessary for an effect, sufficient for an effect, or part of an INUS condition. For probabilistic theories of causality, causes make a difference to the likelihood of their effects. In counterfactual theories, the comparison is to an unactualized possibility where the cause does not occur. The progenitor of difference-making accounts is David Hume, who, in grounding cause-effect relations in regularities, tried to reduce causal connections to patterns of association.

The connection approach, which holds that patterns of difference-making need to be explained by real causal connections, is rooted in the mechanical philosophy of the seventeenth century as articulated by scientists and philosophers like Descartes and Robert Boyle. The key feature of these accounts is that they suppose that cause and effect must either be in direct contact, as when one ball strikes another, or they must be connected by an intervening process or mechanism, as when

DOI: 10.4324/9781003200956-7

a match lights a fuse which burns until it reaches an explosive. The connection tradition sees singular rather than generic causal relations as primary, and while it grants that comparisons that show difference-making may help explain causal reasoning, causation cannot be reduced to difference-making.

The challenge for connection approaches is to say just what these connecting processes or mechanisms are. In this chapter we'll focus on two related approaches to answering this question. Causal process theories try to give a direct answer to what a causal process is that is grounded in contemporary physical theory. New mechanistic theories seek to ground causation in a notion of mechanism drawn from an analysis of scientific practices that extend beyond physics into the life and social sciences.

SALMON'S THEORY

At the conclusion of Chapter 4, we discussed the problem of doing it the hard way—as exemplified by the stories of the improbable hole in one and the miraculous goal line stand. The moral of these stories is that events occur as the result of causal processes that are often irregular or unlikely, and it is the existence of the process rather than the regularity of the association that links cause to effect. This conclusion seems hard to deny, but it provides us little insight without some account of what these causal processes are.

One possibility is that causal processes are simply chains of events. Striking the golf ball (*c*) causes it to fall in the hole (*e*) via a sequence of intermediate events like hitting the branch (c_1), bouncing on the green (c_2) and so on. But this answer seems only to put off the problem, because we still have to ask how each of the links of this chain is causally connected. We can interpolate finer grained events between the links, but then what connects these? We seem stuck in a regress. Reflecting on the problem of causal chains, Wesley Salmon (1980; 1981) became convinced that accounts of causation that treat causation as fundamentally a relation between events were on the wrong track. Rather than trying to reduce processes to chains of events, we need to introduce the notion of a causal process as a fundamental category, with events occurring when these processes interact. Here's a baseball example to illustrate. The batter swings at a pitch, sending the ball towards the outfield where it strikes the wall and bounces to the ground for a double. The swinging bat and the pitched ball are causal processes, and the bat's striking the ball is an event, where these processes interact. That interaction produces changes in each of the processes. The ball changes direction and speed, while the bat is momentarily deformed and slowed by the collision. After the interactions, the ball continues on its altered course until it strikes the wall in another interaction, and so on.

In Salmon's account, both the moving ball and the stationary wall count as causal processes, because both carry persistent causal structure. Physicists have long understood that the distinction between rest and uniform motion is not an absolute one, but can only be judged relative to some frame of reference. It makes sense therefore from a physical point of view to treat them on a par. Salmon does not attempt to define what causal structure is, but gives lots of examples. Sound waves carry voices, radio waves carry signals, a flying arrow carries kinetic energy, and so on. What makes all of these things causal processes is that they **propagate** causal influence through time. Without the propagating sound wave, the words you speak could not influence the actions I take. Without the energy of the arrow, it could not penetrate the target, and so on.

Salmon needed to distinguish between genuine causal processes and what he called pseudo processes, which appear to be processes, but which cannot propagate causal structure. Suppose that a search light is located at the center of a circular domed stadium. As the operator rotates the search light, a spot of white light travels in a circle around the dome. While the beam of light passing from the searchlight to the wall is a causal process, the patch of light moving along the wall is a pseudo process. The moving light patch *looks* like a legitimate process, because the patch continuously moves along the wall as the light rotates, but this is an illusion. The patch itself doesn't carry causal influence, but is just a side effect of the actual causal process.

According to Salmon, what distinguishes causal processes from pseudo processes is their ability to transmit marks. Suppose the spotlight is shining on a fixed point of the dome. If you place a red filter in front of the light, the color of the beam will change. The light beam transmits the mark introduced by the filter to the wall, so the spot on the wall changes as well. But if you begin to rotate the spotlight without moving the filter along with it, the spot on the wall will move, but it will lose its red mark. What this shows you is that the moving patch isn't really a process at all.

While causal processes **transmit** marks, it is causal interactions that **produce** the marks in the first place. These interactions occur whenever causal processes intersect. In the batting example, when the moving bat strikes the moving ball, both the bat and the ball are marked. The shape of each is slightly changed in the collision, as is their momentum. Perhaps other marks, like a smudge of dirt, are introduced as well.

While Salmon's account is motivated by macroscopic physical processes like the baseball example we've been discussing, it can extend to many different kinds of causal processes. Suppose I'm walking my dog and I see another dog walking towards me. My dog is a barker, so I cross the street. Did I interact with the other dog? Strictly speaking, no. When I saw the dog, what really happened was this: sunlight (a causal process) reflects off the dog (a causal interaction), and the reflected light (a causal process marked by the dog) propagates until it strikes my eye (another interaction), triggering further processes and interactions until ultimately my brain activates the neural circuits which move me across the street. I'm causally connected to the dog across the street, but only via the propagating patterns of light.

Let's take stock of the image of causation Salmon's theory gives us. It suggests that the world consists of a vast network of causal processes and interactions. If we think of these causal interactions as events, then we can see that an earlier event is causally connected to a later event just in case there is a pathway forward through this network from the earlier event to the later event. Returning to our story of the hole in one, the reason the sliced shot that hit the tree caused the hole in one is that there exists just such a pathway.

Salmon's theory is a theory of singular causation. It suggests that what stands in causal relations are particular objects, processes and events. If Joe smoked and Joe got cancer, the question of whether his smoking caused his cancer is really a question about whether the various cigarettes he smoked interacted with his body in ways that produced changes to his lungs and other parts of his body that ultimately produced cancers. Whether smoking caused cancer in him does not depend upon the existence or strength of a regular conjunction between smoking and cancer in a population; it is about the actual causal processes and interactions going on inside him. The fact that lung cancer is more frequent in smokers than non-smokers gives us good reason to expect that smoking caused Joe's cancer, but that is an epistemological point. What makes it true, if it is true, that the smoking caused the cancer is that those events are connected via the causal network.

CONSERVED QUANTITY THEORIES

Relatively soon after Salmon published his first account of the process theory, critics convinced him that it needed modification. At issue was his theory of mark transmission. Salmon used the criterion of mark transmission to distinguish causal processes from pseudo processes. A causal process is *capable* of carrying a mark. But to say that it is capable of carrying a mark is to say something not just about what does happen, but what could happen; it refers to counterfactual possibilities. To say, for instance, that the beam of light is capable of transmitting a mark is to say that if we were to mark it by changing its color, that mark would be transmitted.

This was an unwelcome result for Salmon. It suggested that his theory was really just a variety of counterfactual theory, and as a good empiricist, Salmon was suspicious of counterfactuals, whose truth conditions seemed vague at best, and which seem on the face of it to make claims whose truth depends not upon goings on here, but in these shadowy possible worlds. There must, he thought, be some empirically detectable feature in this world which marked certain goings on as causal processes and interactions. Causal facts, he thought, are facts about the actual world. But what facts?

Salmon (1997) ultimately adopted a suggestion by Phil Dowe (1992) that the facts required are facts about conserved quantities. Many of the most basic and important principles of physics are conservation laws. These laws all express the idea that within a closed system, certain quantities are always conserved. Energy, for instance is conserved, even if it changes form. Conservation of energy is what accounts for the constancy of the motion of a pendulum. As the pendulum falls it accelerates towards its midpoint, transforming potential energy to kinetic energy. On the upswing the pendulum gradually slows, converting the kinetic energy to potential energy. This back and forth of the pendulum would go on forever, with the sum of kinetic energy and potential energy remaining constant, except for the fact of friction. Friction, which dissipates energy as heat, slowly reduces the total of kinetic and potential energy, with the result that the pendulum swings just a bit less high each time, until it finally comes to rest. Mass also is a conserved quantity, implying for instance, that in chemical reactions, the total mass of the reactants will be the same at the beginning and end of the reaction. Our understanding of what the conserved quantities are ultimately comes from our physical theories, and these can change. The astute reader may have noticed that mass and energy aren't always conserved. Einstein's famous mass energy equation $E=mc^2$ shows that mass in a system may be destroyed to produce energy. What is properly conserved, according to current physical theory is the combination of mass and energy.

The basic idea of a conserved quantity theory is that (1) causal processes are what conserved quantities persist in, allowing them to propagate through space and time, while (2) causal interactions are intersections between processes where these conserved quantities are exchanged. To illustrate the idea concretely, consider again the baseball example. The baseball and the bat travel towards each other, each characterized by conserved quantities, linear and angular momentum. When they strike in an approximately elastic collision, the linear and angular momentum of each changes—but total momentum is conserved. That's an interaction. The redirected baseball now carries its momentum towards the wall, until it meets the wall, where another interaction and another exchange of conserved quantities occur. The actual situation is complicated both by the fact that the batter is continually altering the swing through his body movements and because bat and ball are also subject to other forces (notably gravitation and air resistance), which mean that their momentum is changing somewhat as they travel through the air. Still, the basic picture seems clear—causal processes allow conserved quantities to persist and be propagated, while causal interactions produce changes via exchanges of these conserved quantities.

Salmon saw important parallels between the view of causality that he had developed and the mechanical philosophy of Descartes and other seventeenth-century philosophers and scientists. Seventeenth-century mechanists understood causation as depending ultimately upon interactions between bits of matter colliding with each other, and it was in this period that the first conservation principles were formulated. The subsequent history of science has shown that momentum is not the only conserved quantity, and that particle collisions are not the only kinds of interactions, but that doesn't alter the basic picture. Salmon (1984) maintained that all causal connections ultimately depend on a small number of "fundamental causal mechanisms" governed by basic physical laws.

EVALUATING PROCESS THEORIES

Salmon's approach to causation has some very attractive features. It offers a scientific answer to what causation ultimately is. It allows us to understand how things that are irregular or improbable can be causes and effects. It is also a reductionist theory in the sense that it tells us how in principle we could translate all talk of causes into something else. And the conserved quantity version of the theory does not depend upon counterfactuals, but refers to measurable quantities in the actual world.

Despite its advantages, Salmon's theory has a significant flaw. In an influential critique of Salmon, Christopher Hitchcock (1995) argued persuasively that neither the mark transmission theory nor the conserved quantity theory gives an adequate account of causal relevance. Causal relevance can be variously defined, but it is a form of difference-making. If some fact or event is causally relevant to an effect, it must make a difference to whether, how or when something occurs or does not occur.

Hitchcock offered examples that show it is quite common for marks to be transmitted, and for conserved quantities to be exchanged, without those events making a difference to some effect. Suppose, for example, that in our baseball scenario there's just a bit of dirt on the bat, and imagine that, when the bat strikes the ball, the ball is marked with a tiny smudge. That marking of the ball with the smudge is a legitimate causal interaction, and the ball carries the mark to its next interaction with the wall. But that mark makes no difference to whether or where the ball hits the wall; the mark is causally irrelevant. In the baseball case, the collision between ball and bat makes (at least) two marks on the ball, changing both its momentum and its coloring. The momentum change, but not the color change is causally relevant to where the ball ultimately lands. A similar story can be told about the conserved quantity version of the theory, since surely when one marks something with a smudge of dirt, conserved quantities are exchanged.

Another way of expressing the causal relevance worry concerns what we might call the level of the interaction. Suppose your child is crying and you give her a hug to calm her. That's an interaction, and doubtless you have exchanged some conserved quantities. If she is warmer than you, thermal energy from her skin might warm your hands. The pressure from your grip slightly compresses the tissue in each of your bodies. But surely none of that explains *why* your hug stopped her crying. You could have hugged her in quite different ways (using a different arm or touching different locations on her body) and could have achieved much the same result. And at the same time, if somebody else who she did not know gave her a physically similar hug, her reaction might have been entirely different. The kind of interaction that makes a difference to the child's behavior is at a different and higher level than that of fundamental physical processes. What is relevant to the child's reaction is not the physical interaction, but the social-emotional-cognitive interaction embodied in your hug. By reducing causal connection to fundamental physical interactions, the causally relevant features of the interaction seem to be left out.

The broader point about conserved quantity theories is that they are difficult to apply in domains outside of physics. We have good reason to believe that raising interest rates curbs inflation or that childhood traumas can cause anxiety and depression—but where are the conserved quantities involved? Presumably all causal processes, including psychological, political, economic and social processes, in some way depend on physical processes and interactions that obey conservation laws, but this fact plays no role in identifying or testing causal hypotheses in these higher-level domains.

In response to these concerns about causal relevance, Salmon conceded that causal connection is not by itself sufficient for causal relevance, but he still maintained that it was necessary for causes to be connected to effects by causal processes. This latter claim though has also been challenged. One difficulty is that many effects appear to be caused not by the occurrence of something, but by a nonoccurence. The cookies burn because I didn't take them out of the oven; a child gets measles because he was not vaccinated; the crop fails because of the drought. In these cases of causation by omission, there is no possibility of a connecting process between cause and effect, since the cause is something that didn't happen. Proponents of process theories may counter this complaint by arguing that omissions aren't really causes. What causes the cookies to burn is the time in the oven, and while it is certainly true that my failure to take them out of the oven explains why they burned, it was the oven, not me, that caused the burning.

A type of case that is more difficult to explain away is called causation by disconnection. Jonathon Schaffer (2000) offers a compelling example. Guns fire when a pin (the striker) hits the primer at the base of the cartridge, igniting the gun powder). In many guns, the striker is propelled towards the cartridge by a spring. The spring is compressed when the gun is cocked, but is held back by a lever called a sear. Pulling the trigger lowers the sear, allowing the cocked spring to release, propelling the pin towards the cartridge to fire the gun. The point to notice here is that the trigger is not itself connected to the spring; its role is to disconnect the sear from the spring, so the spring can cause the gun to fire. This kind of setup is also called double prevention, because the cause prevents or interrupts something that was preventing something else. Schaffer contends that a process theory along the lines of Salmon or Dowe cannot be adequate because requiring that all causes be physically connected to their effects has the absurd consequence that pulling the trigger doesn't cause the gun to fire.

On the other hand, not all cases of double prevention seem like causation. Here's an example: Chris thinks it is a terrible mistake for Alice and Bob to marry, so he rushes to the church to stop the wedding. He's stopped just short of the church by a long train blocking the crossing, and only arrives after the ceremony is complete. The train prevented Chris from preventing the wedding. Does that make the train a cause of the wedding? Most would say not. The puzzle here is why we want to account pulling the trigger as a cause of the gun firing, but we don't want to count the train as causing the wedding.

MECHANISMS

Salmon developed his theory of causation as part of an attempt to understand scientific explanation. On Salmon's view, to explain some phenomenon (event, or process, or whatever) was to situate it in the vast network of processes and interactions that is the causal structure of the world. Most of his work focused on how to explain phenomena by describing the interacting processes that produce the phenomena—in the way we might explain how the ball landed in the hole by tracing the various processes and interactions leading from the shot on the tee to the ball rolling in the cup. He called this aspect of explanation **etiological**—because it explains why a phenomenon occurs by looking at interacting processes in its past. But Salmon recognized a complementary aspect of explanation that he called **constitutive**, which explains what happens *within* a phenomenon. To understand this distinction, consider what it takes to explain a complex event like World War I. The etiological aspect of the explanation will look at the processes leading up to the war, seeking to understand the network of events that caused it. But the constitutive aspect of the explanation will seek to explain how this event plays out. World War I is a massively complex event, and to explain what's going on within it, one needs to understand the actors, processes and interactions involved—what they did and where they did it—and all of the ways these various things are organized and intertwined to constitute that event.

One might try to resist the conclusion that there need be a constitutive aspect to explanation. World War I, you might think, isn't really a single event, but is instead just a big collection of smaller events, and maybe it is just these smaller events that need to be explained. Unfortunately, if you look more closely, the same worry arises for just about any event. Turning the page of this book is, from the point of view of your body, an immensely complicated collection of events, requiring a vast array of cells, organs, and tissues whose activities are carefully orchestrated to allow you to move your fingers and flip the page. There are many more cells involved in this bodily activity than there were soldiers in World War I. Essentially all things and goings on in the world—molecules, cells, ecosystems, births, explosions, extinctions—are wholes made of parts.

In the years since Salmon originally developed his mechanical philosophy, a set of philosophers of science have argued that the key to understanding the constitutive aspect of causal explanation is to understand the structure of composites they call mechanisms. These "new mechanists" (Craver and Tabery 2016; Glennan 2017) think that understanding the concept of mechanism is the key to giving a more scientifically acceptable account of the world's causal structure. The new mechanist understanding of causation and causal explanation stands within Salmon's mechanistic tradition, but by appealing to multi-level mechanisms and constitutive relationships, it seeks to avoid some of the problems that have been raised for the Salmon and Dowe accounts.

You need only open an issue of *Science* or *Nature* to see that a great deal of scientific research is characterized by scientists as attempts to understand the mechanisms responsible for various phenomena. The term "mechanism," in contrast to the term "law of nature" so often discussed by philosophers, is in every scientist's working vocabulary. Nonetheless, it is by no means clear just what scientists mean by the term, and indeed it is likely that scientists in different fields use the term in distinctive ways.

If you look up "mechanism" in most dictionaries, you will find two primary definitions. The first says that a mechanism is a system of interacting parts, typically understood to perform some function. The second says that a mechanism is a process by which some outcome or product is produced. The system sense of mechanism is probably most familiar in technology and engineering. Clocks, cars and power drills are mechanisms in this sense. The processual sense of mechanism is the primary sense in most scientific contexts, for instance when biologists speak of the mechanisms of protein synthesis or of reproduction or of speciation. The two senses are intertwined, but the processual sense is primary. What makes a system a mechanism is that one or more mechanistic processes go on within it. For instance, a cell is a mechanistic system because various of its parts are engaged in mechanistic processes like protein synthesis, cellular metabolism, and photosynthesis.

Proponents of new mechanism have argued that the diverse kinds of mechanisms discussed across the physical, life and social sciences, as well as engineering, all share certain features, which can be summarized succinctly in a definition I call **minimal mechanism**:

> A mechanism for a phenomenon consists of entities (or parts) whose activities and interactions are organized so as to be responsible for the phenomenon
>
> *(Glennan, Illari, and Weber 2022, 145).*

You can use this definition to characterize both systems and processes. For instance, a can opener is a mechanistic system whose phenomenon is opening cans, while photosynthesis is a mechanistic process that converts carbon dioxide and water into oxygen and glucose.

A key feature of this definition is that mechanisms are identified by the phenomenon for which they are responsible. A mechanism can be responsible for some phenomenon in one of two ways; it can either produce the phenomenon or underlie the phenomenon. In the first case, a mechanistic process or system produces some product, and describing the mechanism gives an etiological explanation of what caused the output to be produced. In the second case, the phenomenon is something that is constituted of rather than caused by the phenomenon, so the explanation is a constitutive one. To see the difference between these two kinds of explanation, consider what happens when a muscle contracts. We can ask what **produces** a muscle contraction, and this is most immediately a neural signal that initiates the contraction. We can also ask what **underlies** the muscle contraction. Here we look at the cellular structure of muscles, recognizing that muscle tissues consist of bundles of muscle fibers, which are elongated cells, and that these cells contain in them myofibrils, which in turn are composed of the protein filaments, actin and myosin. When these filaments slide along each other, the myofibrils shorten, thereby shortening the muscle cells and tissues of which they are a part. Strictly speaking it's wrong to say that the sliding of the actin and myosin filaments cause the muscle's contraction, since the sliding of the filaments is not prior to or distinct from the contraction of the muscle. The collective sliding of all of these filaments really **constitutes** the contracting of the muscle.

According to the minimal mechanism definition, mechanisms have two kinds of constituents. First there are the entities or parts. These are what we would ordinarily call objects or things - baseballs, cars, buildings, and such, but also entities that are very much smaller, like cells or proteins, and entities that are very much bigger, like moons, stars or galaxies. They can include living things, like us, and social things, like families or trade unions. These entities are generally composites, composed of parts, so things we call systems or structures also fall within the scope of entities.

Second, mechanisms contain activities and interactions. These are, roughly, whatever it is that entities do, either on their own, or with other entities. Different kinds of entities can engage in different kinds of activities and interactions. Molecules, for instance, can change their shape, and they can bond together. Cells can divide or signal other cells. Animals can do things like run, smell, eat and fight. Some animals like us can talk, and otherwise communicate. There are two words here—activity and interaction—but they form a single category. The word "interaction" implies some kind of going on between or among two or more entities, while "activity" is inclusive of "solo" activities like breathing or walking.

Minimal mechanism makes clear that in order for a mechanism to produce or underlie its phenomenon, the activities and interactions of its constituent entities need to be organized in specific ways. For a heart to pump blood properly, for instance, the electrical signals that coordinate muscle contractions in the various chambers must be precisely timed. And the same constituents, organized in different ways, can produce very different phenomenon. Wiring resistors in line or in parallel, will for instance, generate different patterns of current and resistance in an electrical circuit.

New mechanists are particularly concerned about exploring the relation between causation and constitution in complex systems like animals, ecosystems, or societies, where things are going on at many levels. A representative example from Carl Craver is shown in Figure 9:

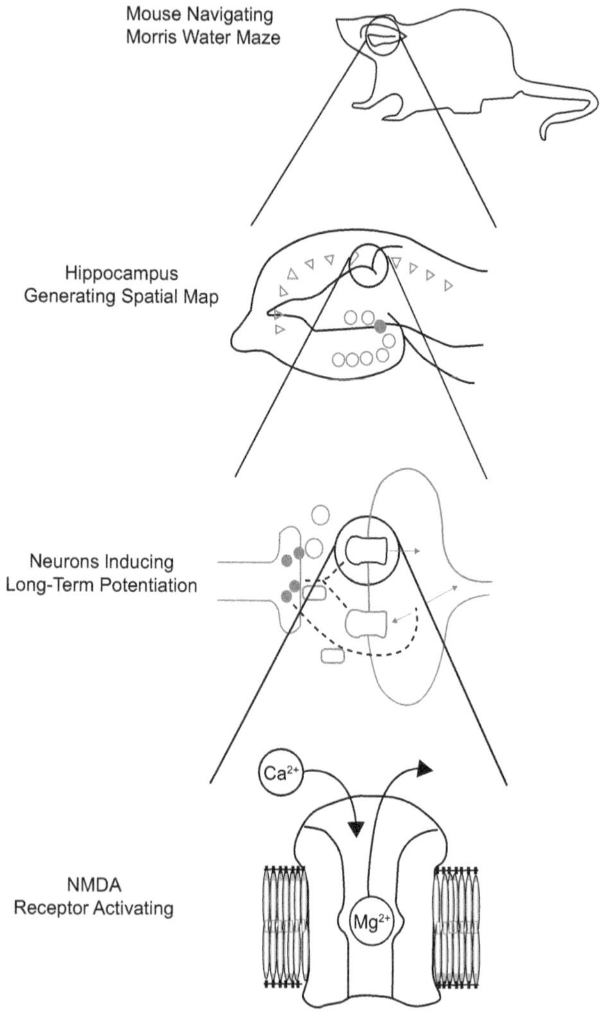

FIGURE 9. Levels of spatial memory from (Craver 2007)

Craver here is describing the mechanisms by which mice (and by extension animals like us) acquire spatial memory. The top-level phenomenon is an activity of a mouse—their navigating a water maze. As the mouse travels through the maze, it relies on a spatial map encoded in its hippocampus. Successful travel through the maze reinforces the map by a mechanism known as long term potentiation—which strengthens synapses between neurons that encode the spatial map. Underlying that mechanism in turn is the activity of an NMDA receptor, which is an ion channel in the neuron's membrane. The relationship between these levels is not causal but constitutive.

It's worth comparing the new mechanist account of causal structure with the one given by Salmon and Dowe. Salmon's processes and interactions correspond roughly with the new mechanists' entities and activities/interactions. Salmon emphasizes that processes are what allow causal structure to persist and says that they include things like baseballs or walls—in short, they are objects or entities. Both Salmon/Dowe and the new mechanists think that these entities interact, but they give rather different accounts of the nature of the interaction. In contrast to Salmon and Dowe, who argue that one particular kind of change occurs when objects interact—namely an exchange of conserved quantities—the new mechanists hold that there are many different kinds of interactions which change the interacting entities in many different ways. Consider again the example of you hugging your child to calm them. This is an interaction which changes both you and the child, but the principal things that it changes are not physical quantities (though doubtless these do change), but rather higher-level features of the complex entities which are you and your child. What changes, most obviously is behavior (the crying stops), but there are many other physiological changes to both of you—in heart rate and blood pressure, levels of stress hormones, and so on. One of the apparent advantages of the new mechanist account over a conserved quantity approach is that it has a better answer to the causal relevance objection. It is the cognitive and emotional features of the hug, rather than any facts about its physical characteristics, which are relevant to explaining its effect on the child's behavior.

Another apparent virtue of the new mechanist account is that it provides a better account of how to deal with causation by disconnection. Schaffer's gun is a mechanism if anything is, and the mechanist can argue that even if there is a (designed) disconnection between two parts of the gun (the sear and the firing pin) when the trigger is pulled, the activity of pulling the trigger is an interaction with the gun that is productive of the firing. Details need to be added here, but the main point is that disconnection is a very common feature of many mechanisms, and the mechanistic framework may allow us to understand how lower-level disconnections may be constitutively relevant to higher level production.

The fact that there are so many different kinds of activities and interactions suggests an important fact about causation. Whenever causal connections occur, they always occur through specific kinds of activities and interactions. Elizabeth Anscombe (1971) was one of the first to make this point, noticing that "cause" is a schema term that we learn only after we learn verbs describing more specific activities. First we learn words like "hit," "scrape," "pull," and "yell" and only then do we come to understand more general causal terms like "cause" and "do."

Anscombe's point is about learning, but it may have a metaphysical consequence. All of the activities named by these verbs turn out to be ways of causing. For instance, I might cause you to turn around either by yelling your name, or by tapping on your shoulder. The word "cause" allows us to speak generally about activities that produce change, but it doesn't imply that causing itself is an activity. Maybe it is just wrong to look for some underlying essence of causal connections, since calling something a cause doesn't tell you what it is, but only something about its role.

An analogy might be helpful here. In Darwin's theory of natural selection, evolution is driven by variations in the fitness of organisms within a population. But what exactly is fitness? The correct answer is that there is no such thing as fitness per se, since fitness is only a way of talking about the differential effects of various organismic traits. For one organism, being tall may make them fitter, while for another it may help to be short. Some may gain fitness by speed, while others get it by stealth or armor, or by various behavioral strategies. Ultimately fitness isn't one thing, but a property that many traits can have, depending upon the context. Perhaps this is what causes are like.

Although new mechanists allow that there are many kinds of activities, they need not treat all of them as basic and unanalyzable. For any given activity or interaction, one can typically identify some kind of mechanism underlying it. Suppose for instance that you call my name. That is an interaction between us, but it is an interaction that is constituted by an array of underlying mechanisms. There are the mechanisms of speech production that require the coordinated activities of various parts of the vocal tract, and all of the neural machinery required to control it. There are also the mechanisms of speech recognition, which involve a variety of acoustic and cognitive mechanisms, and there are of course the mechanisms by which the speech signal—a complex aperiodic compression wave—is transmitted through the air. An interaction like this can be called "mechanism-dependent." On the new mechanist's view, much of the work of describing the world's causal structure comes down to figuring out how activities and interactions depend upon underlying mechanisms.

MECHANISMS AND CAUSATION

The crux of a mechanistic account of causation is the assertion that causes and effects are connected by intervening mechanisms. Here's one way to put it:

> **Mechanistic Causation:** An event *c* causes an event *e* if, and only if, there exists a mechanism by which *c* contributes to the production of *e*.
>
> *(Glennan 2017, 156)*

What makes it the case that your calling my name causes me to turn my head? It is simply that there is an intervening mechanism by which the calling produces the turning.

The mechanistic account requires only that *c* contributes to the production of *e*, rather than producing it outright, because effects generally require a constellation of causes to produce them. It implies, for instance that the grinding of the flour I used to bake the cake was a cause of the baking of the cake, even though a lot more had to happen besides the grinding to produce the cake. In this respect, mechanistic causation follows Lewis's approach of treating anything in a chain of causally relevant events as a cause of an effect.

Like both Lewis and Salmon/Dowe, the mechanistic account takes individual causal relations to be primary. A general causal claim like the claim that turning the heat too high will cause the butter to burn will be true only because it is a generalization over particular instances.

Mechanistic Causation holds that what makes singular causal claims true is *the existence* of a mechanism. Nothing is implied about whether or not we know what the mechanism is. This feature makes sense of the very common situation in which we have good reason to believe there's a causal connection between events, but where we don't know what the mechanism is that connects the events. In the single case, simple contiguity may give us reason to believe there's a causal connection. If, for instance, someone were to take a bite of food and then immediately have an allergic reaction, we would have good reason to believe that something about the food produced the reaction, even if we had no idea of the mechanism by which the reaction was produced. Similarly, in the general case, a regular association may give us strong reason to believe that there is a causal connection even if we don't know what the mechanism is. This is what happened in early lung cancer research, where the patterns of association were sufficiently strong to justify inferring a causal link between smoking and lung cancer, even though the mechanisms of carcinogenesis were not well understood.

Let's compare Mechanistic Causation to some of the other approaches we've discussed. The account is closely related to the Salmon/Dowe view, for on both views mechanistic processes provide the connection between cause and effect. The difference is that new mechanists have a different analysis of what these processes are, emphasizing that interactions are change producers and not trying to reduce them to a single kind of physical interaction, as the conserved quantity theory does.

There are also affinities between the interventionist account and the mechanistic account. Many causal models can be interpreted as models of mechanisms, with the variables representing properties of the parts in the mechanism, and the structural equations representing the effects of the interactions between these parts. Pathways through these models can be interpreted as representing mechanistic processes. At the same time causal models do not in any obvious way introduce levels into causal structure, and it is a matter of some dispute whether one can use interventionist notions to characterize the constitutive relationship between mechanisms and their parts.

The main difference between the two accounts has to do with their differing understandings of counterfactuals. The interventionist account is fundamentally a difference-making account, taking the truth of interventionist counterfactuals as basic facts that ground the truth of causal claims. The new mechanist account, grants that these counterfactuals are true, but insists that the reason why they are true is because of an intervening mechanism. An example will illustrate the point. An interventionist claims that what makes it the case that a certain drug has a certain effect, is that if we were to intervene on the drug—say to withhold it or change its active ingredient—then the effect would be altered. The mechanist agrees, but insists that what makes this counterfactual true is the fact that there's a mechanism connecting cause and effect. If a drug lowers cholesterol, then, all things being equal, withdrawing the drug would raise cholesterol (that's the interventionist counterfactual), but the reason this is true has to do with the mechanism by which the drug's ingredients alter the body's metabolism.

The mechanist has a similar strategy for accounting for causal laws. The mechanist need not deny that there are causal laws, but can insist that what laws do is describe the behavior of mechanisms. Here's an example: Mendel's law of independent assortment is the principle that if a parent has two pairs of genes at different loci (say *Aa* and *Bb*), the probability of an offspring receiving a particular gene at one locus (say *A* versus *a*) is independent of what is happening at the other locus. This implies that the probability of the offspring receiving each of the combinations (*AB*, *Ab*, *aB*, *ab*) are all equal at 25%. This independent assortment principle works most of the time, and there's a mechanism that explains how it does. In most sexually reproducing plants and animals, genes are located on paired chromosomes, and only half of the material from each parent gets put in the sex cell (egg or sperm). The mechanistic process that constructs these sex cells (meiosis) is structured so that it randomly picks which bits of the paired chromosomes go into the sex cell; there's also a mechanism, called crossing over, which flips segments of the paired chromosome before assigning the chromosome to a sex cell. The net result of this is that you will get independent assortment so long as the loci are either on different chromosomes, or if the loci are on the same chromosome but the distance between them is enough that crossing over is likely. When the loci are close together on the same gene, independent assortment fails, and you get a phenomenon called linkage.

Not all laws can be explained this way. Some fundamental laws aren't explained by mechanisms—a point to which I'll return shortly—and there are other laws, like conservation laws in physics, which don't appear to describe cause and effect relationships at all. Still, many causal generalizations used across the sciences and engineering appear to depend upon mechanisms.

WHAT'S AT THE BOTTOM?

The mechanistic account of causation is not without difficulties. The most basic is that, in the views of some, it is not a theory of causation at all. The idea behind this objection is that what mechanisms do is explain how simple causes are wired together into more complicated causes, but it doesn't really tell you what causation is in the first place. Mechanisms, recall, consist of organized activities of, and interactions between, entities. And what is an interaction between entities, if not a causal relationship? Doesn't this make the account circular?

This objection is reminiscent of an objection we already discussed regarding interventionism. There the objection was that it was circular to try to define causation in terms of intervention, since intervention is itself a causal notion. And the mechanist can respond to the circularity objection in a similar way. The supposition of the objection is that the goal of a theory of causation is to reduce causation to something else—causation is just regularity, or causation is just probability raising or counterfactual dependence. But the challenges to various reductionist approaches suggests that perhaps a reductionist account of causation is not possible—because causation is too basic to be reduced to something else. An alternative, non-reductionist approach, seeks to illuminate the connection between certain causal concepts, without trying to eliminate them. For the interventionist, what is illuminated is the connection between causation and intervention, while for the mechanist, what is illuminated is the relation between causation, law and mechanism.

But while the mechanist is not promising a conceptual reduction in which the analysis of the concept of mechanism is given in non-causal terms, there is a different kind of reductive move that the mechanist makes, which may raise further objections. According to the mechanist account, what makes it the case that *c* causes *e* is that there is a mechanism connecting them. This mechanism depends upon causal interactions between the mechanism's parts, but the fact that these interactions are causal depends upon further mechanisms, and so on. Consider a concrete case. Why is it that your calling caused me to turn my head? There's a mechanism underlying your calling my name, which involves, among other things neural signaling within the motor cortex, and the result of these signals are the muscular motions that force air through the vocal cords and mouth so as to produce certain sounds. From there, there are mechanisms for transmitting this sound through the air, and mechanisms for decoding the signal on the other side. But each of these mechanisms depends upon other mechanisms. For instance, the mechanism by which motor neurons initiate muscular changes in the vocal tract depends upon bundles of neurons, signaling each other via action potentials. Down further, there are mechanisms underlying those action potentials.

The looming question is where do these layers of underlying mechanisms end? Perhaps they can go on forever—mechanisms all the way down...—but more likely one reaches a point where there are causal interactions between parts, but where there is no mechanism underlying the interaction. It might for instance be a basic fact that when two electrons get near each other they repel each other. Call such interactions **fundamental interactions**. Critics of the mechanist approach claim that whatever causation really is, it's determined by what's going on at this fundamental level. Maybe at this level, all there is, is regularity, which would imply that the mechanistic theory reduces to a regularity theory, or maybe there are some fundamental laws of the sort proposed by Armstrong, Dretske and Tooley, or maybe things at the bottom are imbued with fundamental powers (Mumford 2009).

One reason to be cautious here is that, given what quantum mechanics tells us about the structure of this fundamental level, it is likely inappropriate to conceive of it consisting of interacting objects (Kuhlmann and Glennan 2014). Exactly how to conceive of fundamental goings on and their relation to goings on at higher levels and larger scales is a subject of ongoing philosophical and scientific research. We will just have to leave the "what's at the bottom?"' question unanswered and recognize that, regardless of how such issues are resolved, developing a full picture of causal structure will require understanding the mechanistic constitution relations that obtain between mechanisms at different levels.

SUGGESTIONS FOR FURTHER READING

Salmon's "Causality: Production and Propagation" (1981) provides a short and clear introduction to his first account of causation. His classic *Scientific Explanation and the Causal Structure of the World* (1984) develops this account in the context of offering a theory of scientific explanation. Dowe's (2000) *Physical Causation* provides an overview of a variety of process theories, including Salmon's, and develops and defends the conserved quantity theory. My *The New Mechanical Philosophy* (Glennan 2017) provides an overview of new mechanism, and offers a defense of a mechanistic theory of causation. Povich and Craver (2018) offer a useful introduction to the idea of levels and its relation to causation. Jon Williamson's two-part introduction to mechanistic theories of causality (2011a; 2011b) provides a useful overview of most of the approaches surveyed in this chapter.

8

CAUSES, PURPOSES, AND AGENCY

The theories of causation considered in the last five chapters have helped us to develop an image of the world's causal structure consisting of a web of interlocking events and processes occurring at different levels and scales. These theories have generally been informed by and meant to make sense of the practices which scientists use to discover and describe this structure.

In this chapter we turn to another use of causal talk which does not fit easily within the scientific image of causal structure, but which is quite important to us in the practical business of going about our lives. We conceive of ourselves as agents who act in the world on the basis of reasons and purposes. These reasons and purposes may incline us to act, but we think nonetheless that under most conditions we act freely, and that we could have acted otherwise than we did. We tend to think of actions as causes of events in the world, but causes for which actors are responsible, and for which they can be praised or blamed.

The main question of this chapter is how we fit *ourselves* and other creatures with agency into the causal structure of the world. Is there a way of understanding agency as part of the natural order of things? Or does it turn out that agency is an illusion, or even that it suggests that the scientific conception of the world's causal structure is somehow seriously off base? These are big questions which could take us far afield from our topic of causation, so in this chapter I will limit myself to marking the

DOI: 10.4324/9781003200956-8

points of intersection between causation and these questions, and suggesting some directions for reconciling these two important perspectives on the world and our place in it.

FINAL CAUSES

Aristotle argued that there are four kinds of causes—material, formal, efficient and final. As discussed in Chapter 2, the contemporary scientific notion of cause most closely resembles Aristotle's efficient cause. Characterizing a thing's matter and form may be important to explaining what it is, but these features do not, in our contemporary sense, seem to cause that thing. Similarly, it seems strange to call final causes, understood as the aim or purpose of a thing, to be genuine causes. One problem is that many things don't seem to have purposes. What is the final cause of a pebble in the road? And even if a thing has a purpose, causes, in the contemporary sense, are events that bring something about, not ends which the thing might serve. A further difficulty is that, as Aristotle's metaphysics was Christianized in the Middle Ages, final causes came to be associated with divine purposes. Within the scholastic tradition, the natural world and everything in it were assumed to serve divine ends. But by the time of the scientific revolution, philosopher/scientists like Descartes and Newton argued that these final causes were at best beyond human ken, and thus could play no role in scientific explanations of the world.

Despite these difficulties, many things in the world do appear to have purposes. In particular, plants and animals have traits that seem exquisitely designed to allow them to survive and thrive in their particular environments, and this close fit between organisms and their environment seems to demand a causal explanation. The question is how these purposes could count as causes.

One possible way to account for the purposiveness of things in nature is by analogy to the way we account for the purposiveness of the artifacts humans create. When humans build or make something—a bridge, a house, even a work of art—they do so to achieve some end. The builder's end is part of a straightforward causal explanation of why the thing with a certain purpose came to exist as it is. If the final cause of the bridge is to transport goods across the river, the builder's desire to achieve this purpose can account causally for the bridge's being built with the properties it has.

Advocates of natural theology used an analogy to human design and engineering to argue for the existence of God. The theologian William Paley (1743–1805) observed that if we were walking in a field and came upon a watch, we would, seeing its obvious intricacies of design, infer immediately that it was the work of a watchmaker. He then argued by analogy that when we observe all the intricacies of design that enable living things to survive and thrive in their environments, we should immediately infer the presence of a designer—the divine watchmaker.

Design arguments have a number of difficulties, some of them explored by Hume in his *Dialogs Concerning Natural Religion*. Perhaps most obviously, we have known watchmakers and have seen them build watches, but we don't have any experience of divine watchmakers creating worlds. Still, advocates of design arguments are right to point out that design in nature demands an explanation, and maybe for a while, theirs was the best explanation going.

The design argument though was undermined by Darwin's theory of evolution by natural selection, which is able to account for the appearance of design without positing a designer. Populations of organisms exhibit variation, which arises from mutation, sexual reproduction, and other factors. Organisms whose traits better allow them to survive and reproduce will be more successful in transmitting those traits to future generations, leading over many generations to a population whose traits are well adapted to their environment.

While the principle of natural selection can explain the presence of a trait by reference to its adaptive purpose, it does so in a way that is perfectly consistent with rejecting the view that final causes direct things toward an end. Natural selection processes do not involve design, because the variations that give rise to it are not deliberately chosen. Neither does natural selection imply that evolution is progressive, or that current species are more advanced than past species. Nature does not have purposes. Natural selection pushes populations towards packages of traits which fit their immediate environment, but there's no guarantee that what is adaptive now will remain adaptive as the environment changes.

The biologist Ernst Mayr (1961) has suggested that rather than thinking of biological causes as either efficient or final, we should think of them as proximate or ultimate. The proximate causes of an organism's behavior are a combination of the organism's intrinsic physiological traits and processes and the immediate environment with which it interacts. The ultimate causes of an organism's behavior lie in the evolutionary history of the population to which the organism belongs. It is that historical process which gives the organism the genetic endowment, which, under appropriate conditions, allows it to develop its intrinsic physiological traits and processes. Both proximate and ultimate causes explain current effects in terms of past causes, but on very different time scales.

REASONS AND AGENCY

At first glance it seems both necessary and harmless to include reasons as part of the causal structure of the world. It is quite common for us to explain our own or others' actions as being caused by reasons. We might say that Jenny studied for the test because she wanted to get a better grade or Felix hid under the bed because he was afraid of the house guests. Reasons, in the broad way I am using the term, may include all sorts of attitudes—beliefs, desires, fears, goals, and so on.

Philosophers understand reasons to be the things on the basis of which **agents** choose **actions**. Human beings (at least adult ones) are paradigms of agents, and many cognitively advanced animals also exhibit agency. Actions are distinguished from ordinary events on the grounds that actions are done for reasons, while ordinary events are not. My blinking and breathing, being automatic, are events but not actions. Similarly, if someone pushes me into the water, then the falling was an event I participated in, but not an action I took; only if I had jumped in the water of my own volition would it have been an action.

The simplest way to understand actions is to treat them as events that are caused by reasons, but some philosophers have argued that the relationship between reasons and actions cannot be a causal one. One reason for their skepticism has to do with human freedom and responsibility. As Leibniz famously put the matter, "reasons incline but do not necessitate." What exactly Leibniz meant by this claim is a matter of some dispute, but at least one thing he seems to be insisting on is that reasons are logical justifications for actions, rather than events which cause actions. Leibniz also insisted that since reasons do not necessitate, it is possible when someone acts on the basis of reasons that they could have done otherwise than they did. This is what makes us free, and thus responsible for our actions.

The worry then is that if reasons are woven into the causal fabric, genuine agency is lost. As Thomas Nagel puts it:

> [S]omething in the idea of agency is incompatible with actions being events, or people being things. But as the external determinants of what someone has done are gradually exposed, in their effect on consequences, character, and choice itself, it becomes gradually clear that actions are events and people things. Eventually nothing remains which can be ascribed to the responsible self, and we are left with nothing but a portion of the larger sequence of events, which can be deplored or celebrated, but not blamed or praised
>
> *(Nagel 2012, 37).*

There are roughly four ways to respond to Nagel's worry. One approach is simply to accept that our sense of agency is an illusion. A second is to take the fact of our agency as implying that there is something wrong or incomplete about the scientific image of causal structure. A third is to argue that agents and actions are not part of the world's causal structure at all. A fourth is to deny the incompatibility that Nagel alleges, and try to find an account of agency that is consistent with our evolving scientific understanding of causal structure. We'll consider each of these options in turn.

The first response commonly goes under the name of **hard determinism**. It combines the view that the world has a deterministic causal structure with Nagel's claim that free will and agency are incompatible with determinism. It implies that all of our actions, however free they may feel to us, are in fact simply events within the world's causal structure. They may express our desires, values and dispositions, but these things too are ultimately the product of interactions between genes and the environment. And ultimately, none of this is up to us.

While some find hard determinism deeply troubling, there is, ironically, something liberating about it. It still allows us, as Nagel says, to "deplore or celebrate" events, but it leaves out the praise or blame. If we discover that a person who committed a crime has grown up in an environment of poverty, abuse and neglect and we are inclined to view them as a victim of their circumstances, we are expressing a hard determinist sentiment. We may believe in the need to punish the person—say to protect the community, to deter others from action, or to help reform their criminal behavior—but we don't punish out of a sense of retribution. Similarly, when someone does something we find valuable or extraordinary—from Martin Luther King leading the march on Washington to

LeBron James breaking the NBA scoring record, we may celebrate the event and be grateful for the talent, perseverance and courage it took to pull it off, but we still may see these traits as the endowment of fortune, rather than the result of some personal agency.

In my experience teaching, most people are not entirely able to give up on the idea of agency in interpreting the goings on in their own lives, or in the lives of others. Anyone who attends carefully to the world will recognize that a great deal of what we are initially inclined to praise or blame people for depends upon causal factors beyond their control; but most maintain that there is some core bit of the self and the will that maintains control over action. Certainly luck makes it much harder for some than others to do the right thing, but, the thought goes, what we ultimately do is up to us. The hard determinist though insists that even a stripped-down notion of agency is illusion. There simply is no place to put it within the world's causal structure.

Finally, I should mention one avenue of escape that is probably not available to save agency. While the traditional discussion of free will has been framed as a conflict between free will and determinism, as we discussed in Chapter 4, our best contemporary physical theories are indeterministic in important respects, so it may turn out that causal determinism is false. But this fact likely does not provide an avenue for agency because, according to these same theories, what is not causally determined is a matter of chance. It seems very difficult to understand how an agent could be any more responsible for chance events than they are for events caused by things beyond their control. It would be hard, for instance, to blame somebody for an act, if it turned out that their choice to act depended ultimately upon which way an electron bounced in some indeterministic interaction. You can't save agency by chance.

The second response to Nagel's dilemma is commonly called **libertarianism**. (This is a metaphysical sense of "libertarianism" which is distinct from the political sense.) The libertarian takes it as self-evident that we are agents, and argues that we know from experience that our actions are neither causally determined nor the product of chance. Agents can initiate events in the world, as for instance, when I decide to raise my hand; but my decision as an agent is not an event in the world's causal structure. It is, in essence, an uncaused cause.

Some philosophers have adopted this position to accommodate their theological views. An agent's special status as uncaused cause is an attribute medieval philosophers attributed to God, as the supposed creator of the universe, and it was often thought that a person's ability to act on their free will gave sense to the Biblical claim that humans are created in God's image. But if we are not moved by such considerations, it is hard to see how our subjective experience of freedom is sufficient to guarantee that there is no underlying objective causal structure that accounts for why we make the choices we make.

The libertarian position faces major difficulties. One problem is that it is mysterious how the agent could insert itself into the world's causal structure. The most straightforward way to understand what an agent might be is as a Cartesian immaterial soul. This soul would have to be influenced but not determined by goings on in the world, and its acts of will must initiate actions within the world. But, as we discussed briefly in Chapter 2, any account of where and how such interactions occur is elusive at best, incoherent at worst. There are perhaps other ways to understand the ontology of agents—but absent a clear account, it's hard to credit the libertarian position.

A second difficulty has to do with the boundaries of agency. Agency seems to come in degrees and varieties, and it seems to be something that can come and go. We take cognitively and emotionally competent human adults to be exemplars of agents, but what of children, of infants, or those with varying degrees of cognitive impairment? Similarly, many non-human animals—especially our near cousins among species—show signs of agency. They appear to perceive, to acquire and test beliefs, to make decisions on the basis of their own goals, and so on. Do these animals count as agents, and if they do, how far from our neighborhood in the tree of life does this special property extend? And given the fact that all creatures on the planet belong to species that have evolved over billions of years from non-living things, we need to have an account of agency that allows for it to evolve. Finally there is the possibility of artificial agency. Many androids familiar from science fiction would count as agents by any reasonable standard, and while we do not yet have systems that exhibit all these abilities, technologies are rapidly changing, and there are already many robots and software systems that exhibit various features of agency.

A third response to Nagel's dilemma is to argue that it arises from the mistaken assumption that agents and reasons belong in the world's causal structure in the first place. Kant's solution to the problem of free will appears to fall within this category of response. Recall from Chapter 2 that Kant saw the world of things in themselves (noumena) as distinct from the world we experience (phenomena). Kant thought moreover that the concepts we use to organize experience—notably space, time and causation—are imposed by us to make possible experience of the phenomenal world. At the same time, Kant held that genuine agency and acts of the will are noumenal rather than phenomenal. They do not exist in time and space, and they have no causes and effects. If this is right, this preserves agents, action and freedom, as noumenal, while keeping causation in its place in the phenomenal world. Thus, the problem of how to integrate them does not even arise. Kant's solution is clever, but it requires us to accept his argument that the world of experience is essentially a construction of our own minds. And Kant doesn't have an entirely satisfying account of how, if this is so, different agents can live in and interact within a shared world.

The fourth response to Nagel's dilemma, **soft determinism**, holds that determinism or something like it is true, but offers an account of freedom and agency that is compatible with it. Soft determinists look for a naturalistic account of agency, in which agents really are just a special variety of things and actions just a special variety of events. There are many variants of soft determinism, so I'll only sketch some basic outlines.

At its most basic, the soft determinist holds that an agent is free to the extent that they can do what they want, which is to say that they can act on their own will. Freedom in this sense is primarily opposed to constraint. In this most basic sense we are clearly free, as are many animals. What is required to be a free agent in this sense is to have beliefs, desires, goals, and other cognitive states, and for these states to cause actions.

This is a pretty minimal sense of agency, and it seems insufficient for purposes of assigning moral responsibility. If a lion, say, harmed a child, it might be acting on its own desires, but we should not blame the lion, in a moral sense, for the harm that occurs. For that, one needs a more complex psychological structure. Harry Frankfurt (1971), for instance, suggests that the key requirement for being a free person is to have effective second order desires. Frankfurt argues that one of the characteristics of our psychology is that we can have conflicting desires, and that we can also have desires about desires—second order desires. I might for instance desire to eat the last cookie, and also desire to be polite by not eating the cookie. I might want the cookie very much (first order desire), but wish that I didn't want the cookie (second order desire). For Frankfurt the key to being free is to be able to choose among first order desires to make one our will. If I am able to resist my desire to eat the cookie, say because I want to let my son have it, I am exercising free will. If this line of thought is correct, then free will may be unique to humans among animals, and it will be the case that not all humans have it. For instance, a drug addict who wants to quit, but is unable, is not free. They can act on their desire for drugs, but they cannot act on their second order desire to stop their desire for drugs from motivating their action.

You do not need to endorse the details of Frankfurt's analysis to see the value of the general strategy. Agency, on accounts such as Frankfurt's, is a matter of having a certain kind of cognitive structure, one that causally interacts with the world to produce events that count as free actions. This puts agency firmly within the world's causal structure, as it is scientifically understood. One feature of this kind of account, for better or for worse, is that the cognitive structures of agents are determined by their causal histories. This implies that that agents don't get to choose to be free. They either have the good fortune (if you think it good) to have freedom, or they do not. Agents like us, if we are free, are not radically self-determined. If you think that freedom you only have by luck is not true freedom, then you will fall back into hard determinism.

Stepping back from questions of human agency and freedom, it's possible to identify two important features that are required for any entity to count as an agent. The first is **control**. "Control" is a term used in many academic disciplines, including biology, psychology, sociology and engineering. At its most basic controllers are things that can change the state or dynamics of systems. A steering wheel or accelerator control the motion of cars; thermostats control furnaces; central nervous systems control animal movements, and so on. For something to be a controller, it must be able to both detect and change the states of the system it controls. Steering wheels and accelerators must be controlled by human drivers or computers with sensors; thermostats must detect temperature changes; central nervous systems must be able to detect changes to states of the body and the environment. Control systems are systems that control other systems by sensing and maintaining/changing their states. It is plausible to think that any agent, natural or artificial, must be a control system.

The second feature that would seem to be required for agency is **autonomy**. Autonomy is, in general, the ability of a system to act without the control of an external controller. In engineering contexts, we are familiar with the idea of self-driving cars (autonomous vehicles), autonomous drones, robots, etc. Autonomy always comes in degrees. A car may for instance have limited autonomy to stay in lanes and start and stop, greater autonomy to self-drive along a designated route, or even greater autonomy to change a route. Autonomy in varying degrees is a characteristic of animals, and of living things in general. As human agents we deeply value autonomy for ourselves, and in various social roles different actors (parents and children, supervisors and employees; teachers and students; officers and soldiers) are afforded different levels of autonomy. Every agent's autonomy is limited in some ways, but for any agency at all, a measure of autonomy is necessary.

Any naturalistic account of agency will need to illuminate what is distinctive about agency, while simultaneously allowing that agents are things that fit naturally into the world's causal structure. One useful way to think about how both these requirements can be reconciled is by analogy to Daniel Dennett's account of the intentional stance (Dennett 1989). Intentionality is, according to many philosophers of mind, the distinctive mark of the mental. What makes a particular state of mind intentional is that it is about or directed toward something in the world. The attitudes I have collectively called reasons (beliefs, desires, goals, etc.) are all intentional.

Dennett's account of the intentional stance is meant to tell us when we should attribute intentional states to things in the world. His answer, simply put, is that something is an intentional system to the extent that attributing intentional states to it can help us to predict how the system behaves. Dennett identifies three distinct ways we can regard any object or system, which he calls the physical stance, the design stance, and the intentional stance. The physical stance regards the system as a composite of physical parts operating according to physical laws. The design stance regards the system as having a purpose or function. The intentional stance regards the system as having beliefs, desires and other intentional states. Strictly speaking, we can apply any stance to anything we like, but for different systems, the different stances can be radically more or less helpful.

Consider for instance a rock. Taking the physical stance, I can explain why it is just sitting there by appealing to basic physical principles and properties of the rock. The rock stays where it is because it is heavy, because the ground is relatively flat, and because there is enough friction between the ground and the object to keep it in place. I could also take the design stance, regarding the rock as a paperweight. It might indeed be a good paperweight, but regarding it this way wouldn't help me understand any of its features (in the way say that regarding the device in my kitchen drawer as a can opener could). I could even take the intentional stance towards the rock, and explain its immobility by the fact that it just loves the spot it is in and doesn't want to move—but such an explanation fails to add anything to an explanation given from the physical stance.

For people though, the intentional stance is exceedingly valuable. I invite a speaker from across the country to my university, and remarkably she arrives where the talk is to occur just a couple of hours before it is to begin. She came to Indiana via a complex sequence of conveyances—a train, a couple of flights and a cab. How can I explain her appearance? From the physical stance it's essentially impossible. She did indeed get here on the basis of a series of physical movements of her body—including embarking and disembarking from the various conveyances—but certainly I cannot track every one of those movements and trace them back to the neurological and other bodily causes that got her there. The only practical way to predict and explain what she did is to attribute intentional states to her: She read and understood my invitation, she decided to come, she made travel arrangements, and she executed them.

What's interesting about the intentional stance is that you can profitably apply it to systems that you might believe don't "really" have beliefs and desires at all. For instance, if you are playing a computer game against an AI, in order to predict how the AI will behave you will inevitably attribute to it goals and strategies. Your avatar might be hiding in some cave, and you will hope that the bad guy AI "doesn't know you are there." Or you might make some move—a feint—hoping to trick it. And this kind of strategy works. It allows you to explain and predict what the AI is doing without knowing anything about how it is built. We might still think that the AI doesn't really have beliefs, but, from Dennett's point of view, the reality of intentional states consists simply in the presence of real patterns of behavior that can be economically described by attributing these states to things.

The point of this analogy is to show how we might think of agency as a kind of stance we take when describing the behavior of certain physical systems. We might be able in principle to avoid all this agency talk and simply describe all the goings on within this system as embedded within the world's physical causal structure. But this is a practical impossibility, so agency talk is effectively ineliminable. And when agency attributions lead to effective explanation and prediction of behaviors of things in the world—whether humans, animals, social organizations, computers, or whatever—we can take these agents and their reasoning as parts of the world's causal structure.

MENTAL CAUSATION

Our last question in this chapter concerns how to fit minds into the world's causal structure. We have already begun to explore this question in our account of agency. Agents and minds aren't quite the same thing, but they are conceptually intertwined. We are agents and we have minds; we understand reasoning to be something that goes on in our minds that is essential to our agency. Still, what is most important to agency is autonomy and control, while what is most important to mental states is conscious awareness, and it seems possible for things without conscious awareness to have autonomy and control, and things with conscious awareness to lack autonomy and control.

What then are minds? Descartes and other seventeenth century philosophers interpreted this as a question about substances. Dualists like Descartes held that minds were immaterial substances, while materialists, or physicalists as they are now more commonly called, held that minds were material things, differing from other material things only in the way in which their material parts were organized. Contemporary dualists have mostly shied away from substance dualism, and have instead argued for property dualism—a view which holds that certain physical things (most often brains) may have non-physical properties. What these non-physical properties are is notoriously hard to pin down, but they are generally understood to be phenomenal properties or "raw feels"—the painfulness of pain, the redness of apples, the scent of coffee, and so on.

Dualism of either variety faces a challenge concerning mental causation. How can things and their properties, if they are non-physical, make a causal difference in the physical world? The basic answers philosophers have considered—interactionism, epiphenomenalism, and parallelism—are pictured schematically in Figure 10. Mental events are denoted by *m*. Physical events are denoted by *p*. Causal interactions are denoted by arrows. Dotted arrows denote apparent interactions.

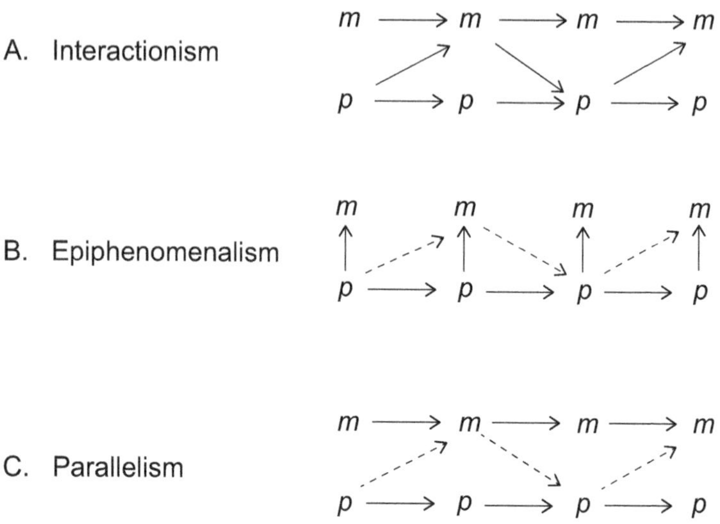

FIGURE 10. Theories of mind-body interaction

Interactionism, the view favored by Descartes, seems to encapsulate our commonsense understanding of the relation between mental and physical. Causal processes flow from physical to mental and back. For example, a bear walks into view (physical event). I see the bear (mental event), which causes fear (mental event), and a decision to run (mental event) which starts my body rapidly moving (physical event).

Mental events and physical events certainly appear to make a difference to each other, and so qualify as causes of each other, but there are two major difficulties that arise if mental events are not also physical events. The first difficulty is that it is not clear where or how these interactions could take place. Mental events do not seem to have locations. If your thoughts aren't really events going on in your brain, it makes no sense to say they are happening between your ears. Also, mental events do not have the kind of properties that would allow them to interact with physical things. We know why a hammer can pound a nail, because it can strike the nail with appropriate force—but how can an event like a "decision" push the body to do something? The second and related difficulty is that we have good reasons to believe that the physical world is causally complete. This means that every event in the physical world (to the extent it has a cause at all) has a complete set of physical causes. Suppose for instance that I move my arm. There is, we expect, a causally complete story, going however far back you want, of the neurological causes of this motion. This story, being complete, leaves no room for additional mental causes.

If there is no room for mental causes to enter into the physical world, we must consider alternatives. According to the first, epiphenomenalism, physical events can cause mental events, but mental events cannot cause physical events. The epiphenomenalist account, as pictured in 10b, envisions mental events as projections of physical events. Strings of mental events are like movies, projected onto our mind's eye. The strings of mental events look to be causally connected to each other and to the world, but all the real activity is going on in the physical world.

By granting that mental events don't cause things in the physical world, the epiphenomenalist avoids violating the principle that the physical world is causally complete. The one-way causal connection also provides an explanation of how our mental life is coordinated with goings on in the material world. But epiphenomenalism is still puzzling. It has no account of how or where physical-to-mental causal connections occur, and it requires us to accept that our experience of our minds making a difference in the world is an illusion.

The third approach, parallelism, holds that mental and physical events do not interact at all. They are on two parallel tracks, dancing together, but never coming into contact. Since parallelists deny any interaction between mental and physical, they need not explain how and where such interactions occur. What the parallelist does need to explain is how, in the absence of interaction, we get the close coordination between the mental and physical. In the ordinary world, we explain coordination by interaction. Even in the case of the non-touching dancers, we explain their synchronized movements by the fact that they are watching each other, or perhaps they are listening to the same music and thus dancing to the same beat.

How do we explain this synchronization without such interactions? The answer favored by Leibniz was divine harmony. A common metaphor to explain this harmony is synchronized clocks. If two synchronized clocks are allowed to run, you will be able to read one clock and thereby tell what the other clock will say. You have coordination without interaction. Of course real clocks lose time, so they will need to interact from time to time to keep in sync; but if nature is made by God, the divine clockmaker, this would not be needed. Parallelism solves the mental causation problem by doing away with mental causation, and the best explanation it can give for the illusion of causal interaction appeals to divine harmony. These are far from ideal moves for someone seeking a naturalistic understanding of the world's causal structure.

It has seemed to many philosophers that the absence of a coherent dualist account of mind-body interaction is reason enough to reject dualism outright. If we can simply identify the mental and the physical, taking mental states to just be certain kinds of physical states, then interaction between mental and physical states becomes unproblematic. When I see the bear, that's a physical event, and so it can be causally related to my beginning to run, another physical event.

The physicalist though, if rejecting dualism, needs some account of what minds and mental states are, that will explain why some things, like ourselves, have mental states, while other things, like the rock lying on the ground, do not. What emerged in the second half of the twentieth century as an answer to this question is functionalism. Functionalists argued that mental states could be identified by their functions or causal roles. For example, pains are the kinds of states which are brought about by injuries to or malfunctions of the body, and which cause avoidance behaviors and efforts to repair damage. When for instance, you touch a hot stove, pain is the state that causes you to withdraw your hand and to rush to put it under cold water. Beliefs and desires can also be understood functionally. Your behavior of pulling an umbrella out of the closet is a consequence of your belief that it might rain, and your desire not to get wet.

When you characterize a state functionally, you are characterizing it by its role (what it does) rather than by its physical properties (what it is). This is crucial for the functionalist story in part because it allows for physically different things to occupy the same functional role. As a simple example, consider the devices that remove corks from wine bottles. There are various kinds of corkscrews, including waiter's corkscrews and winged corkscrews. There are also two-pronged cork removers where no screw is inserted, but instead two thin metal prongs are inserted on opposite sides of the cork. More exotic cork removers include sabers, which can remove champagne corks by striking the bottle's lip, and heated port tongs, which can be used to remove the entire top of a bottle's neck. Philosophers call this phenomenon, where a functional role can be realized by distinct physical mechanisms, multiple realization.

If mental states are functional states, it is possible that they could be multiply realized, and we have evidence in the natural world that something like multiple realization in fact occurs. For example, cephalopods (squids and octopuses), are invertebrates that are phylogenetically and physiologically extremely different from intelligent primates, but which appear to have many of the cognitive traits that primates have. They can learn, they can plan, they can hunt and set traps, and so on. This implies that they can be mentally similar to primates like us while being physically and physiologically quite different. At a finer grained level, we likely will find that multiple realization even occurs within species. Depression, for instance, involves a characteristic set of behaviors and dispositions, but it is unlikely that the neurological processes that underlie depression are the same in all cases.

The idea of multiple realization dovetails nicely with the idea of mechanistic levels discussed in the previous chapter. We can think of the mechanism underlying some behavior as the realizer of the behavior, and we can allow that distinct mechanisms can, on different occasions, underlie the same kind of behavior. Because the underlying mechanisms are diverse, it isn't possible to reduce higher level mental phenomenon to some single physical basis. The resulting position is often called nonreductive physicalism.

Nonreductive physicalism seemed to many philosophers and psychologists to offer the best of both worlds. It endorsed physicalism, holding that each particular mental state ultimately depends upon the physical, while at the same time stressing that mental properties and states are distinct from physical properties and states. Nonreductive physicalism seems to shed light on mental causation. Mental states can cause things in the world, as they are in fact defined by their causal role, but because these mental states are realized by physical states, the causal powers are not something over and above the physical.

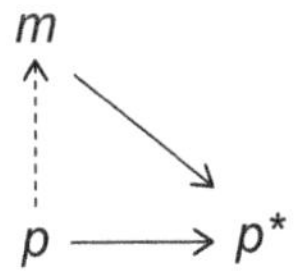

FIGURE 11. Causal Exclusion

This happy state of affairs was disturbed by the philosopher Jaegwon Kim (1989), who offered a simple but compelling argument, called the causal exclusion argument. Kim's argument, schematically represented in Figure 11, is simply this. Suppose we think that mental event *m* causes physical event p^*. Event *m* though is realized by physical event *p*. But physical event *p*, the realizer, is causally sufficient for p^*, so there is no distinctive causal power for *m*. Figure 11 looks like a case of causal overdetermination, where both the *m* and *p* are causally sufficient for p^*, but Kim's causal exclusion principle asserts that genuine overdetermination can only occur when two causes are distinct (as when two assassins both shoot at their target). Given that *m* is realized by *p*, it isn't distinct. The correct conclusion, Kim thinks is that *m* must be epiphenomenal. Kim sometimes calls this argument "Descartes' revenge"—because the retreat to physicalism that was supposed to save the causal efficacy of the mental turns out to make the mental epiphenomenal.

While Kim formulated this argument about the mental, it seems to generalize to rob causal efficacy from any causal interaction occurring above the level of fundamental physics. If, for instance, I strike a hammer with a nail, the hammer only is able to move the nail because of the hardness and mass of the hammer's surface, and this hardness is realized by the molecular structure of the iron of which the hammer is made. The exclusion argument seems, absurdly, to imply that hammering doesn't really cause nails to be driven.

This apparent consequence of the exclusion argument suggests to many philosophers that something must be wrong with it, but the diagnoses of what is wrong vary. One possible answer is to insist that at the individual level, the relation between the realizer and the realized is really an identity relation, which is to say that the realizer and the realized are one and the same, just described in different ways. And if *m* and *p* are identical, then if *p* is causally efficacious, so too is *m*. Whatever the answer, the causal exclusion problem points to the importance of developing an account about how causal relationships obtaining at different levels within complex systems like cells and organisms are connected to each other.

In a famous paper, Wilfred Sellars (1963) argued that philosophy's most central challenge is to resolve the tensions between what he called the manifest and scientific images of man. We see in these debates about minds, reasons, and agency exactly the tension that concerned Sellars. How do we hold together our understanding of the manifest world in which we act with the vast and impersonal world that our scientific investigations have revealed to us? We cannot delve further into these issues here, but it is important to mark them, because questions about causation seem ineliminable on both sides of this divide.

SUGGESTIONS FOR FURTHER READING

Hume's *Dialogues Concerning Natural Religion* (1998) provides a readable introduction to the classical debate about design arguments. Allen and Neal's "Teleological Notions in Biology" (2020) offers an overview of recent work in philosophy of biology on function and purpose.

Paul's *Philosophy of Action* (2022) provides a useful overview of the relation between actions, agents, events and causes. Two useful introductions to the free will problem from very different perspectives are Dennett's *Elbow Room* (1984) and Kane's *A Contemporary Introduction to Free Will* (2005). Two introductory texts in philosophy of mind that highlight the problem of mental causation are Kim's *Mind in a Physical World* (1998) and Heil's *Philosophy of Mind: A Contemporary Introduction* (1998).

9

FINDING CAUSES

Through much of this book our focus has been on what causes are, and what makes various kinds of causal claims true. In other words, our focus has been on conceptual and metaphysical questions. But reasonable answers to these questions have always had to be informed by our understanding of how we are able to acquire causal knowledge. In this chapter we will bring these epistemological questions to the forefront.

Sometimes causal knowledge is easy and immediate. I know from experience that hanging the laundry out causes it to dry, and I know immediately after it happens that the scrape on my knee was caused by the fall from my bike. But in other cases, it is far from clear what causes what. Consider these questions:

- What causes AIDS?
- What causes Black mothers in the US to die in childbirth at ten times the rate of White mothers?
- What effect does masking have on reducing rates of Covid-19 transmission?
- What caused the failure of Lehman Brothers Bank in 2008?
- What caused the crash of Lion Air flight 610 in October of 2018?
- What caused the 2021 insurrection at the US Capital?

All of these are important questions, and none of them have been easy to figure out. The questions above are of two different kinds. The first three

DOI: 10.4324/9781003200956-9

are about generic causation—what causes what?—while the last three are questions about singular causation—what caused that? As we explore techniques for causal discovery and inference, we'll see that the methods for answering these questions are somewhat different.

There will always be more than one good answer to the question of what causes what. Here, for instance are three legitimate answers to the question about the causes of AIDS:

- AIDS is caused by HIV retrovirus
- AIDS is caused by unprotected sex, IV drug use with shared needles, blood transfusions, and transmission from infected pregnant mothers to their children
- AIDS was caused by transmission of, and subsequent mutation of, a simian retrovirus transferred from chimpanzees to humans through consumption of chimpanzee meat

The first of these cites the virus which degrades the human immune system and allows for AIDS symptoms to occur. The second of these identifies the mechanisms by which HIV is most commonly transmitted from an infected person to a non-infected person, and the third identifies the causal origin of the HIV virus in humans. Notice that the first two answers are general causal claims about what *causes* AIDS while the third one is a singular causal claim about what *caused* AIDS. This example illustrates that answering the question "what caused what?" really involves two steps. First it involves discovering facts about the relevant causal structures, and second it involves selecting those features of those structures that best answer a particular question about causes in a given context. The first, the **causal discovery problem**, is the chief concern of this chapter, while the second, the **causal selection problem**, is the focus of the next chapter.

We ultimately acquire all causal knowledge from two sources—observation and experiment. Observation is a matter of watching, measuring, and counting the goings on in the world, while experiment adds the extra step of intervening in the world. First we do something, and then we observe what happens. In this chapter I'll begin by considering what observation is, and what kind of causal knowledge can be acquired from it. I'll then explore the special role of experiment in acquiring causal knowledge and say something of the different varieties of experiments, and of their limits. The next two sections will explore methods of discovery and inference in more complex circumstances, first introducing methods that can be used to infer causal structure from statistical data, and then discussing methods that assemble diverse bodies of causal evidence to make causal inferences. We'll conclude with singular causation, and the special challenges of discerning the causes of things past.

OBSERVATION

Hume's regularity theory of causation begins from an argument about how we acquire causal knowledge. Hume argued that we are never able to *perceive* a causal connection in a single case. His justification for this claim is that we cannot ever, on first observing an event, infer what effect will follow: if we haven't ever put our head below water, we won't know that we won't be able to breath; if we have never touched a hot stove, we won't know that it will burn. Hume thought that this fact about our experience was enough to show that we cannot *perceive* causal connections directly, and that we only *infer* causal connections after we have become habituated by a regular association.

Hume's argument is dubious. It's not at all clear that our not being able to predict an effect implies we can't perceive a connection between cause and effect. If I'd never put my head in water, and somebody threw me in, would I not immediately, as I struggled not to drown, perceive that the water was preventing me from breathing? If I stepped outside on a very windy day, would I not feel the wind push me as I struggled against it? If I put a piece of laundry on the clothesline, would I not see it pulling the line downward?

That Hume could deny these apparently obvious cases of causal perception is best explained by his theory of perception. Hume thought that our perceptions were constructed out of bundles of what he called impressions—patterns of light, dark, and color in our visual field, sensations of hardness and resistance to our touch, feelings of warmth or cold, scents, tastes, and so on. These impressions are what later philosophers would come to call sense data. Hume claimed that if we search among our impressions, we can find nothing like an impression of connection.

Maybe we can't, but by this standard, neither can we have impressions of objects, or of the spatial and temporal contiguity that is required for the regularity theory. If you consider what, in the ordinary sense of experience, you perceive, they are objects—like your laptop or phone, or a tree or your dog or a person. You also perceive what these objects are doing. The dog might be barking, or the computer might be flashing.

You also can perceive the press of your Mom's hug squeezing you and warming you. What you don't perceive, in the sense of ordinary conscious experience, is sense data. When you look at the table across the room, you see the table. You don't see a set of colored regions and lines, and as you move around the table, even though the patterns of light hitting your eye change, your perception of the table remains constant.

Hume imagined, as a bit of speculative perceptual psychology, that the things we perceive are constructed from sense data, and indeed something like this appears to be right, but this hypothetical construction is not conscious reasoning or inference. Sense data is not something we can perceive. We *infer* that our sensory experience is constructed from sense data on the basis of our beliefs about how our perceptual systems work.

Hume's argument that it is impossible to perceive causal connections seems then to be on shaky ground. But even if Hume is wrong, it certainly won't be the case that every causal connection can be perceived. There are causal connections that allow us to speak over cell phones, that account for the efficacy of drugs, that explain why sea levels are rising, and so on. We cannot directly perceive many of the events involved in these cases, much less the causal connection between these events, so we still need an account of how we can infer the existence of unseen connections like these.

It will clarify matters if we are more careful about what we mean by observation and perception. Perception is a cognitive capacity that we and many other animals have to detect, by means of our senses, goings on in our immediate surroundings. We can perceive what philosophers like to call "ordinary middle-sized objects" by vision and touch. We can perceive voices and other sounds; we can also perceive the position and configuration of our own bodies (proprioception). Observation, on the other hand, is chiefly a scientific concept. It refers to our ability to detect and measure by means either of our senses or of instruments. The point that observation often requires instruments is of key importance. When I look through a microscope, I am able to observe the walls of a cell as well as the nucleus and other cellular bodies. But it would be wrong to say that I *perceive* the cell. This is even more clear if we consider Geiger counters. Particles emitted from nuclei of decaying atoms register on the Geiger counter as clicks, but it's wrong to say that we are hearing the particles, as they make no sound.

Scientists routinely construct devices to allow us to observe a world beyond perception—radio telescopes, CT scanners, ultrasound machines, gene sequencing devices, neutrino detectors, and so on. These devices provide causal links between the imperceptible and perceptible worlds—images on computer screens, readings on dials, and so on. We cannot define causation in terms of regularities in what we observe, because our observations depend on these causal links.

Contrary to what Hume thought, it does seem possible at least sometimes to perceive causal connections immediately. I don't need to infer from repeated experiences that the water I have fallen into now is preventing me from breathing. Sometimes too, we may, with the help of instruments, be able to observe causal connections between events that are beyond our perceptual powers. A surgeon, for instance, may be able to read a CT scan to see a burst blood vessel that caused a stroke.

But sometimes is not always. Often we must infer the existence of entities, events and causal connections that we cannot perceive either directly through our senses or indirectly by means of our observational instruments. One familiar example is infectious diseases. We can observe that diseases—flu, measles, Covid, AIDS, etc.—appear in people who come into contact with each other. From the pattern of contact we infer that the disease agent is somehow transferred from one person to the other. We can't see the disease agent or the act of transmitting the disease, and we are often uncertain about just what kind of contact is required, but the patterns of outbreak give us very good reason to believe there is a causal connection. We are essentially utilizing Mill's methods of agreement and difference.

Observed correlations are not enough of course, since correlation alone does not imply causation. Consider, for example, cholera. In cholera outbreaks you will observe patterns of spread that suggest transmission between individuals. In fact, though, cholera is not transmitted by contact with people, but by ingesting water or food that contains cholera bacteria. An outbreak in a household is not caused by contact among the household members, but via the common cause of their shared water and food sources. People can sometimes be linked together in causal chains of cholera transmission, but this occurs principally in cases where feces from infected persons contaminate water consumed by non-infected persons. Sorting out genuine causes from non-causal correlations is the most important challenge to causal discovery, and an essential tool for this task is experiment.

EXPERIMENT

Experiment is a foundational piece of scientific method, but it is, before that, a tool that all humans use in everyday life. To experiment, at its most elemental, is to do something, and then to watch what happens. Babies experiment when they put things in their mouths, or when they push objects off the trays of their highchairs. We experiment when we start pushing the buttons on our phones to try to figure out how to make a call, change a setting, or open an app. We can experiment with each other when we tell a joke, flirt, get angry, or offer a gift. We are doing something in the world, and then seeing what happens. While correlations are patterns of happenings in the world, experiments are interventions that can make things happen.

Here is a kind of experimental reasoning with which we are all familiar.: Suppose a light goes out, and we wish to figure out why. First, we flip the switch, to make sure it is switched on. If that doesn't work, we might plug another light into the same outlet to check and see if power is coming to the outlet. If the power comes on in the other light, the problem may be the light bulb, so we get a new one. If the other light didn't work either, there may be some problem with the circuit, so we go switch the breaker. Perhaps the problem is with the power to the house. We flip on other switches to see if we have power anywhere. If not, we look outside to see if our neighbors' lights are on, and so on.

This kind of diagnostic reasoning is used by all sorts of people—engineers, plumbers, doctors, farmers, and cooks. The various diagnostics allow us to ferret out the causal structure of a given situation. Success with this kind of reasoning often depends upon expertise. When the mechanic diagnoses what's wrong with your car, they don't know immediately, and must perform tests, but they know enough about the causal structure of cars to know what tests to perform. Often this knowledge comes in the form of awareness of something like INUS conditions. The point of the tests is to figure out which required conditions are absent (and thus preventing an expected effect) or which of the various possible conditions are occurring which jointly explain the effect. Background knowledge is crucial in guiding experiments.

There are many varieties of experimental reasoning used both by us in our ordinary lives and by scientists as they seek to understand the world's causal structure. Some basic classifications will help us survey the territory. First, we contrast experimental reasoning in the broad sense, which is something everyone uses to understand the world and make things happen in it, with scientific experimentation, which involves a variety of techniques used by experts to discern aspects of the world's causal structure. My discussion so far has been about experimentation in the broad sense. Second, within the category of scientific experimentation we can distinguish between deterministic experiments, which seek to uncover deterministic causal dependencies, and probabilistic experiments, which seek to uncover probabilistic causal dependencies. The line between these two varieties of experiments is by no means precise, but this classification will be helpful in describing different kinds of experimental designs and the different kinds of causal information they convey.

An experiment is an intervention into some chunk of causal structure. If that structure is deterministic, a single trial should reveal the effect, and repeated trials should replicate the effect. Recall an experiment you might have done in a high school chemistry class. Take some gas and enclose it in a sealed vessel of volume V, with a thermometer and a pressure gage attached. After measuring initial pressure and temperature, turn on a burner under the vessel. After a while, measure the new temperature and pressure. You should detect that the pressure has increased in proportion to the temperature increase. This is the Boyle-Charles law—$PV=kT$.

Although the law is expressed as an equation, with no causal direction, in experimenting on this system, we interpret it causally. We can represent this experiment using Pearl's Do operator:

$$\mathrm{Do}(T=t) \rightarrow P=kt/v$$

The Do operator represents the intervention, and the formula says that intervening to set the T to t causes the pressure P to become kt/v. T is the independent variable, P is the dependent variable, and the relation revealed by this experiment is causal dependence.

This is a deterministic experiment, in the sense that a given setting of T (and V) should in principle always yield the same value of P. The reality though is never this simple, since there are inevitably errors or limitations of precision in the interventions and measurements in actual experiments. Nonetheless, were you to conduct experiments in a lab, taking repeated measurements at different temperatures and then graphing the relation between T and P, you would get something very close to a straight line.

Generalizations like the Boyle-Charles law hold widely, but they are seldom truly universal. Their crucial property is that they are invariant under interventions. You can intervene to set temperature, pressure and volume to a wide range of different values, and the general relationship between the variables will remain the same. Invariance in causal generalizations always has its limits. In this case, if you heat the vessel too much you might melt it, or blow it up. And the gas law only works for gases, so if you cool the contents of the vessel enough to change it to a liquid state, the invariant relation breaks down. One thing you can learn from experiments is information about the range of invariance of a causal generalization.

Many engineering tests have the form of simple deterministic experiments. Build a device that is supposed to respond in systematic ways to interventions, and conduct an experiment to see if it behaves as designed. For instance, ignite a rocket motor and measure the thrust produced. Or test the control system in an aircraft, moving the stick and rudder pedals, and measure changes of pitch, yaw, and roll.

The causal connection between the intervention and the effect obtains because of a connecting mechanism. When the invariant causal relations break down, the explanation is generally a breakdown in the mechanism. If pulling the stick up fails to raise the plane's nose, this might occur for a number of reasons. Maybe the mechanical linkage between the stick and the control surfaces on the tail has been damaged; alternatively, the plane may be in a stall or not have sufficient forward momentum to allow those control surfaces to alter the flight path. One can apply similar tests to organisms and other naturally occurring systems. Exposing a mammalian eye to bright light will cause its pupil to constrict. This is a more or less deterministic response of the ocular mechanism—though it too can be interfered with or broken down, as when a doctor gives eye drops that maintain dilation.

A crucial requirement for all experiments is control. To understand the effect of an experimental intervention, scientists make efforts to eliminate or hold constant the effects of variables that could confound the effects of the intervention. The aim is to approximate as much as possible the conditions for an ideal intervention like that specified in Chapter 6. For instance, to test the effect of temperature on pressure, it is important to use a vessel that will not expand as pressure increases; or to measure the rate of response of a plane to flight controls, one will want to measure responsiveness under standardized atmospheric conditions. Experiments can also be conducted to establish the range of invariance. Once basic causal dependencies have been established under controlled standard conditions, tests are performed under expanded sets of conditions to explore how far one can deviate from these conditions before things break down. In testing an aircraft, for instance, one wants to understand at what speeds or angles the aircraft will stall.

Deterministic experiments can reveal causal dependencies in conditions where the causal structure is sufficiently fixed, so that one gets, at least approximately "same cause, same effect." To figure out which lights a light switch turns on, I only really need to experiment once, because the wiring of the house is fixed, and presuming relevant background conditions (like no power outages), the switch should operate the same way on every occasion.

Probabilistic experiments, on the other hand, are required where the presence or size of the effect of an intervention is not constant across individual cases. In such circumstances, one can only estimate the probability of an effect, or the expected value of an effect. Variations in effects across cases arise, for the most part, because there are variations in unmeasured properties of the individuals involved in them. If we plant multiple seeds of the same variety, we do not expect them to grow into identical plants, or indeed that all of them will germinate. If we give multiple patients with the same condition the same drug, we do not expect them to respond in the same way to treatment. We expect variations in these cases, because we know each situation is different. Seeds will not be genetically identical, and they may have developed in different ways, owing to variations in their parental plants and their environments, and despite our efforts to control it, there will be variations in the environment in which we plant them. Similarly, the patients to whom we administer a drug will, because of their differing origins and histories, vary in manifold ways—genetic, anatomical, physiological, etc., and they too will live in different environments. Given such variations, experimentation typically must be conducted on large populations, and the effects will have to be described by statistical measures.

As a first example of a probabilistic experiment, consider the use of animal studies to establish whether environmental chemicals cause cancer. The basic design of such studies involves exposing groups of animals—most often mice or rats—to varying levels of the potential carcinogen, and then performing diagnostic tests on the animals to detect tumors. The results of the experiment can be measured in various ways—the rate of occurrence of tumors within various experimental groups, the number or size of tumors, the time required for tumor formation, and so on. In evaluating what such experiments reveal, there are two major questions. The first, called internal validity, concerns whether the experiment has correctly measured the effect within an experimental population. The second, called external validity, concerns whether the results found in the experiment apply to individuals outside the experimental population.

Internal validity can largely be established by proper experimental design. The main requirement is to design the experiment so that any observed effect will not likely be due to confounding variables, measurement errors, or chance. A first step is to use an experimental population large enough to make it unlikely that chance alone can account for variations detected. Another important step is to control for possible genetic causes by using genetically uniform animal species. A third step is to compare background rates of incidence of the effect with the effect in the experimental group. Mice and rats develop cancers in their lungs, breasts, etc., just as humans do, and to establish that a chemical is carcinogenic, researchers must establish that exposure to the chemical raises the rate of these cancers over the background rate within the population as a whole.

External validity is a trickier matter. Researchers use animal models to study carcinogenicity for both ethical and practical reasons. Ethically, we cannot expose people to potentially carcinogenic chemicals simply to investigate their effects. Practically, animals like mice are relatively cheap to produce and house, they can be bred to create genetically uniform strains, and they can be experimented upon and euthanized if necessary. Despite these advantages, it is unclear when we can use the results of animal studies to infer things about causes in human populations. While researchers choose the animal models they do because of their substantial similarities to humans with respect to the phenomena under study, differences always remain. Moreover, the conditions of the experimental intervention do not match the conditions of natural exposure in human populations. For instance, to establish a carcinogenic effect, researchers may expose their animals to far higher levels of carcinogens than humans are typically exposed to. Does a high incidence in cancer in those animals imply that there will be a real if smaller effect on humans exposed to lower doses? The result is suggestive, but hardly definitive. The limitations of animal studies explain why medical researchers testing the efficacy of potential treatments must resort to clinical trials. To understand whether some drug or therapy is effective in people, we must test it on people.

Much clinical research is conducted using an experimental model called a randomized controlled trial (RCT), and RCTs are widely used across the life and social sciences. They are with good reason considered the gold standard for establishing the existence of causal connections between an independent and a dependent variable, but their success depends upon careful design, and one must take care in interpreting their results.

To illustrate how RCTs work, consider the design of trials used to test the efficacy of Covid-19 vaccines. In an RCT, the experiment randomly assigns members of their experimental population into two groups—the experimental group and the control group. The experimental group receives the treatment (here, the vaccine), while the control group receives no treatment, or is treated with an inert placebo. The researchers then track the health records of the two groups to see if there are differences of effect in potential dependent variables. In the Covid-19 example, for instance, one would measure the frequency of positive Covid-19 tests in each population, as well as, for those who get Covid, measurements of the duration and intensity of the disease.

The reason that randomization is important to these experiments is that it eliminates the effect of uncontrolled variations between experimental and control groups. The presence, absence, or intensity of an effect will depend upon many things besides the independent variable. In the case of Covid-19, for example, susceptibility to Covid-19 will depend upon a host of individual and environmental factors—genetics, other health conditions, age, diet, occupation, living circumstances, to name just a few. We know, for instance, that Covid-19 is transmissible between people, so we understand that people living or working in more crowded conditions will be more likely to be exposed to the disease. Randomization does not eliminate these variations, but it helps assure that they are spread evenly across the experimental and control groups, so that effects of these variations on the dependent variable will, on average, be the same across both groups. It is likely, for instance, that accountants, nurses, and food service workers will be differentially affected by Covid-19, owing to differential workplace exposure, but, with sufficient study size, randomization should allot such individuals approximately evenly between experimental and control groups, so that the difference measured between the two groups can be reasonably attributed to the experimental intervention. Sometimes, in cases where background knowledge suggests known confounding variables, experimenters will deliberately balance groups across them. For instance, they might assure that the gender composition or age distribution of the experimental and control groups are balanced.

As with other experiments, the aim of the experimental design in an RCT is to make the intervention on the experimental variable as close to ideal as possible. This fact explains why in many clinical drug trials the control group is administered a placebo. Health outcomes are often affected by psychological mechanisms. Most simply, a patient's belief that they have received an effective treatment may improve their condition, regardless of the physiological efficacy of that treatment. Administering a placebo will, it is hoped, screen off this effect. If subjects in both groups cannot tell whether they've received the experimental intervention, their psychological states cannot systematically explain any difference in outcomes. Similar considerations explain why many clinical trials are double blinded. In a double-blind experiment, neither the patient nor the clinician know whether the patient is in the experimental or control group. This protocol guards against circumstances in which accidental or deliberate differences in behavior towards the members of the two groups might make a difference to the dependent variable. Collectively, these features of design in RCTs are meant to assure the internal validity of the experiment; that is, they help to assure that if there is a difference in effect between the experimental and control groups, that difference can be attributed to the causal efficacy of the treatment.

While these examples have been drawn from medical research, RCTs and other experimental methods have application across the sciences. One can use variants of these methods to test the effectiveness of fertilizers on crop yields, the effectiveness of teaching methods on student learning, the influence of training on implicit bias, or the effect of tax policies on savings and investment.

Perhaps the most important limitation in all of these methods is external validity. Can one take a study that establishes an effect in an experimental population and apply it in "the wild." Will a drug that proved effective in reducing anxiety among 20-30 year old men, prove equally effective with women, or children, or the elderly? Can a study done comparing two pedagogical techniques in one school apply to other schools in the district, or in the country, or in other parts of the world?

Can a tax incentive that proved effective in promoting investment in France work equally well in England or Norway? Our confidence in answering these questions will depend in part on how much we know of the mechanisms relating cause to effect, but often they will only be answered by further experiment.

Observation is contrasted to experiment, because in experiments the fact that differences appear after interventions gives us reason to think that those interventions were what accounted for the differences. Sometimes, however, we are presented with a situation very much like an experiment, but where no experimenter was required to engage in the intervention. These are called **natural experiments**. A classic example occurred during a cholera outbreak in London in 1854. The physician John Snow hypothesized that cholera was transmitted by contaminated water, and undertook to examine the water sources used by residents affected by the outbreak. Snow discovered that the risk of contracting cholera was highly correlated with the water company from which one's residential water came. Residents receiving water from companies whose water source was upstream of sewage discharges into the Thames were relatively unaffected by cholera, while those receiving water from companies whose water source was downstream of the sewage discharges were severely affected. Equally important, a resident's water company was largely uncorrelated with other variables that might affect the disease. For instance, houses next door to each other were sometimes supplied by different water companies, and the same water companies supplied both rich and poor, men and women, the elderly and children. In short, the conditions were very similar to a randomized controlled trial, in which, in effect, residents were "randomly" assigned to a water company, and where the resulting variation in disease incidence could be measured.

The Covid-19 lockdowns of 2020 generated a variety of conditions that could be called natural experiments. The lockdowns produced massive reductions in travel and commuting, which impacted emissions of CO_2 and other pollutants. By looking at changes in local and global measures of air quality and other climate variables, one can find evidence for causal connections between emissions and these variables. The lockdowns also produced massive social changes, providing a wealth of opportunities to assess phenomena like the impacts of remote work or distance learning on performance, or the social implications of isolation or spending larger amounts of time with family. Differences in approaches to public health policy across jurisdictions also created natural experiments, allowing comparisons between the effect of different interventions like mask or vaccine mandates or stay-at-home orders. These "experiments" are natural in the sense that the different conditions are not the result of deliberate scientific intervention, but they are hardly controlled, and observed correlations cannot automatically be taken to be causal. Germany and Sweden, for instance, had very different initial approaches to lockdowns, and different patterns of Covid-19 incidence, but there are enough other variations between the social and political context, as well as in the patterns of disease spread, to make simple inferences about the effects of different policies difficult. More generally, the set of behavioral and social changes wrought by the pandemic were so numerous and various that there are limits to what we can do in untangling their effects.

FINDING CAUSES IN STATISTICAL DATA

Valuable as experiments are, there are many cases where practical or moral considerations prevent us from using them to test hypotheses about causal structure. This is especially true of the causal structure in large scale physical, biological or social systems and processes. We cannot perform repeated experiments on the earth to measure the effects of changing carbon emissions on climate. Neither can we in any straightforward way use experiments to identify the causes of health care disparities in the United States, or to assess the effects of incarceration of parents on the economic welfare of their children. We can do small localized experiments to explore possible mechanisms— for instance demonstrating that sealed vessels containing greenhouse gases retain more heat than vessels with ordinary air, or demonstrating how health care workers may exhibit implicit bias in performing certain tasks. But such experiments do not tell us if and how these mechanisms combine with other mechanisms and conditions at larger scales to produce the effects we are trying to understand. Developing a reasonable account of these complex causal structures demands deployment of a variety of methods and sources of evidence.

Within the past thirty years or so, statisticians, computer scientists and philosophers have begun to develop techniques for testing models of causal structure against observational data. In a sense, these techniques are methods from moving from correlations to causation, without requiring experiment. These methods employ the causal modeling techniques we briefly introduced in Chapter 7. The details of how these methods work are technical, but we can say something here about the problems they are trying to solve, as well as their limitations.

To illustrate the issues involved, let's consider what causes gender gaps in reading and mathematics performance among children. Historically over the last decades boys have tended to outperform girls in mathematics, and girls to outperform boys in reading—though there are considerable variations across countries and times (Marks 2008). What might cause these differences? One possibility is that there are real biological differences between boys and girls that account for this differential performance, but social explanations seem much more likely, both because cross-cultural variations in these numbers are quite significant and because there are few plausible mechanisms by which primary and secondary sexual characteristics could make a difference in these varieties of cognitive performance.

Let us then imagine the kind of data collection and analysis that could provide insights into the social causes of the gender gap. The first step is to identify and measure a set of variables that past research and background knowledge suggest could be causally relevant to performance. Some of these will be connected directly or indirectly to gender, while others may not. Some factors that plausibly matter are:

- Homework: the amount of time the student spends on homework
- Engagement: the degree to which the student actively participates in class activities
- Level: the level of the highest math class the student takes before being assessed.

These factors in turn will be impacted by other variables. For the sake of illustration, let's add, in addition to the student's Gender, their socioeconomic status (SES) and their parent's educational level. An additional variable, Advising, reflects a measure of how aggressively a student is advised to pursue mathematics. We can put these together in a directed acyclic graph (DAG), with arrows representing hypothesized causal dependencies (Figure 12).

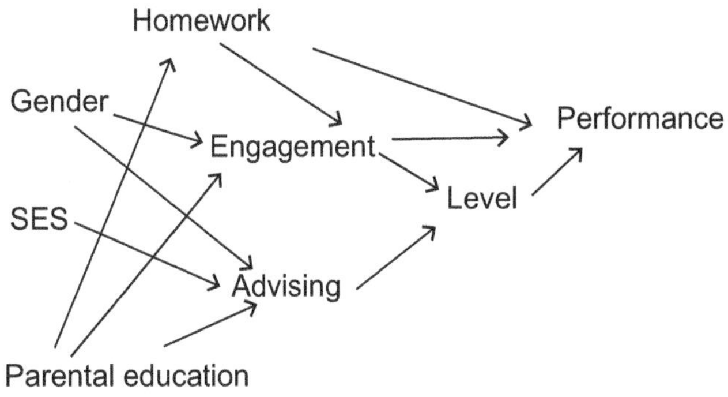

FIGURE 12. Factors influencing test performance

This model represents a hypothesis about the causal structure underlying student performance. It contains assumptions about dependence relations between variables, suggesting, for example, that gender has no direct influence on performance or homework, but that it does have a direct impact on classroom engagement and advising. These assumptions may or may not be true, and need to be tested.

It is possible to test such a model if we make an important but plausible assumption about causal structure. This assumption, called the Causal Markov Condition, is that the values of variables should statistically depend only upon their direct causes (i.e,, those variables shown as their parents within the graph). So, for instance, the value of Level, should depend only upon Engagement and Advising. Another way to put this is to say that direct causes screen off remote causes (like Gender and SES) from their effects (see Chapter 4). If it turns out that the Causal Markov Condition is not satisfied by the observed data, then we can infer that the hypothesized causal model is incomplete or incorrect. If, for instance, it turned out that the value of Engagement was not independent of SES, then the model would need to be revised.

Statisticians and computer scientists have developed tools that allow researchers to test the degree to which a proposed causal model predicts the observed data. Additionally, they have developed causal discovery algorithms that, given a set of variables, can identify a set of possible models (i.e., possible DAGs) that would account for the patterns observed in the data. In general, there are multiple models consistent with any given body of data—a phenomenon called underdetermination—and researchers may need additional data to choose among the possible models.

It bears repeating that causal models are abstract, idealized and partial representations of causal structure, always defined with respect to a variable set. Different variable sets will lead to different causal models. The set of variables chosen will be based upon background knowledge of what factors might be causally relevant, and they will also be heavily influenced by what can be measured. It is very common though that the values of variables that have been measured are in fact influenced by other unmeasured **latent variables**. Consider as an example what educators like to call the growth mindset. Math teachers understand that a crucial factor in a student's success is their belief that their mathematical aptitude is not fixed, but can grow as a result of their efforts. This psychological disposition might be a latent variable that would causally influence both a student's willingness to do homework and their willingness to engage in class. Causal modeling algorithms can be used both to test causal models with unmeasured latent variables and to propose models with such variables as common causes to account for otherwise unexplained correlations.

While DAGs are powerful tools, they cannot represent all aspects of causal structure. One difficulty mentioned in Chapter 7 concerned representing feedback loops, and we can see that this could be a concern in this case. Consider the relationship between Homework, Engagement and Level. The graph in Figure 12 represents Homework as a fixed quantity, whose only parent in the model is SES. Homework is a direct cause of Engagement, reflecting the idea that working more on homework will enable and encourage students to engage more in class. It is clear though that patterns of homework and engagement may change over time and will be mutually reinforcing. Additionally, homework and engagement may not only affect the level of classes taken, but the level of classes may create conditions that encourage or require more homework and engagement.

Causal models, no matter how well they account for observed data, are subject to the same concerns with external validity that we encountered with interpreting experiments. When a causal model is inferred from or tested by a set of data, it will at best capture the causal structure that gave rise to that particular set of data. There is no guarantee that a similar causal structure will be found elsewhere. This is evident in the case of the gender gap. Because both educational systems and gender roles and dynamics vary widely across communities and times, there is no reason to expect that we will see similar gender gaps in different places;

indeed in the last few years, the gender gap between girls and boys in mathematics in the US has largely been eliminated (Reardon et al. 2018) and ongoing research will need to account for growing populations of children not fitting into cisgendered categories. Moreover, even in cases where the kinds of gaps are similar, it is unlikely that the causal explanations of those gaps and the interventions we might use to address them, will be similar across those cases.

PUTTING TOGETHER THE EVIDENCE

Causal discovery can be easy when effects are immediate and uniform, but gets much harder when an effect is not immediately discernible or if the effect is intermittent. The connection between smoking and lung cancer is a classic example. Smoking is neither necessary nor sufficient for getting lung cancer. Some smokers don't get cancer, and some non-smokers do. Moreover, when a smoker does get cancer, its onset can be decades after the smoking began, and even decades after the smoker quit. In the face of such challenges, determining with certainty whether an association implies a causal link depends upon a diverse body of evidence.

Within public health and medicine, the kinds of evidence required to establish causality have been codified in the Bradford Hill Criteria. These criteria were proposed in 1965 by the English epidemiologist, Sir Austin Bradford Hill (1965). Bradford Hill, along with Richard Doll, were among the first to provide evidence for the link between smoking and lung cancer in 1950. The strength of association reported by Bradford Hill and others strongly suggested a causal link, though a disinformation campaign by allies of the tobacco industry was enough to keep the public in doubt well into the 1960s (Oreskes and Conway 2010). Writing in 1965, Bradford Hill proposed nine criteria for determining whether an association was causal. These criteria were inspired by Bradford Hill's experience in the debates over smoking, but they can be applied generally to any kind of association between variables *C* and *E* within a population. The criteria were as follows:

1. Strength of Association: How much more likely is *E* given *C*?
2. Consistency: Is the association between *C* and *E* observed in different circumstances and times, and by different researchers?
3. Specificity: Is *C* associated specifically with *E*, and with nothing else? For instance, is smoking associated only with lung cancer, and

not with other cancers? And is *C* the only potential cause strongly associated with *E*?

4. Temporality: Does *C* precede *E* in time?
5. Biological Gradient: Does the amount or level of *C* increase the risk or level of *E*? For instance, is heavier smoking associated with greater risk of lung cancer?
6. Plausibility: Is the idea of a causal connection between *C* and *E* plausible given our understanding of potential mechanisms connecting them?
7. Coherence: Does the hypothesized causal association cohere with other known facts about the effect.
8. Experiment: Is there direct experimental evidence that manipulating *C* can change *E*?
9. Analogy: Is the hypothesized relationship analogous to other known causal relationships?

Bradford Hill's criteria should be viewed as heuristics for guiding research and assessment rather than as a strict list of individually necessary and jointly sufficient conditions. Probably the only must-have is temporality, since causes must precede their effects. Some causal associations will be stronger than others, and some may be localized to particular populations or circumstances. Not all causes are specific. Also notice that the criteria aren't entirely independent. In particular, plausibility, coherence and analogy are closely related.

Bradford Hill's criteria were formulated to establish causes of disease, but the principles, or something close to them, are widely applicable to many other patterns of causal association. We could use them, for instance, to assess links between poverty and crime, or between interest rate changes and unemployment.

A second example of a technique to put together a body of evidence in support of a causal relationship can be found within the practice of evidence-based medicine (EBM). The EBM movement is an effort within the medical community to make sure that clinical interventions on patients are causally efficacious. Evidence-based approaches, it is thought, will serve both the patient and the larger community by assuring effective treatment for patients and by cutting down on wasteful and ineffective procedures.

Advocates of EBM argue that medical evidence comes in a hierarchy, represented in the pyramid in Figure 13. At the bottom of this pyramid lies background information and expert opinion. In the view of EBM advocates, this kind of evidence by itself is of little use. Such information may help guide research, but is no replacement for evidence garnered by actually studying the association between some intervention and its effect as it occurs in actual patients. Next on the pyramid come various kinds of observational studies—either studies of individual cases (i.e., individual patients and their course of disease) or observational studies comparing, for instance, those who engage in some behavior or have some risk factor, from those who don't. Suppose for instance we were interested in the impact of exercise schedules on depression. In an observational study we would collect data on some population's exercise habits and correlate them with frequency or intensity of depressive symptoms. Above such studies in the hierarchy is the randomized controlled trial, where we intervene on a treatment group and compare them to a control group. We might for instance, impose an exercise routine on the treatment group, while letting the control group continue with their existing exercise patterns. But individual studies, especially small ones, are hardly sufficient to guarantee the efficacy of a treatment. For this reason, we are interested in systematic reviews and other forms of meta-analysis which collect and analyze the results of multiple studies on related phenomena in different populations. With this highest quality of evidence, we have good reason to be confident in the causal efficacy of the treatment.

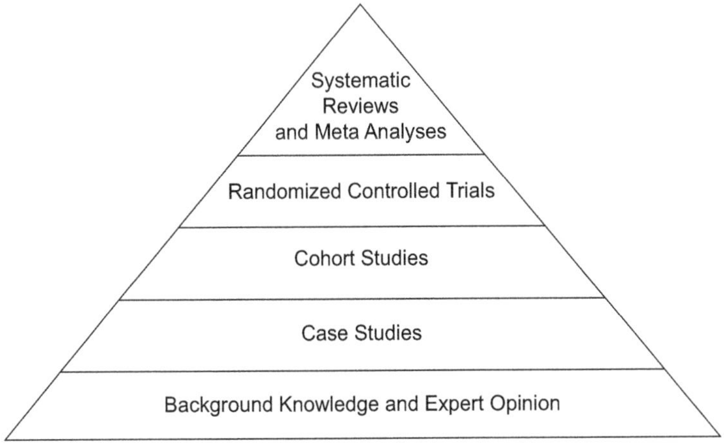

FIGURE 13. An evidence-based medicine hierarchy.

My quick presentation of EBM glosses over many controversies in the medical community and among philosophers of medicine about exactly how this pyramid can be constructed, or indeed whether evidence can be arranged into such a strict hierarchy. Some critics raise concerns that background information, and particularly basic clinical research on, e.g., pharmacological mechanisms, may be underrated. Others point to limitations in the design of various kinds of RCTs. Despite these disagreements, there is no disputing the value of the pluralistic approach to gathering evidence concerning the causal efficacy of possible interventions in medicine. Moreover, analogous evidence-based approaches have gained adherents in many other fields, including education, public policy, and policing.

LOOKING BACKWARD FOR ACTUAL CAUSES

Our focus in this chapter has chiefly been on finding generic causes: What accounts for gender gaps in math performance? Can we be sure that smoking causes cancer? But a rather different type of investigation is needed for finding actual causes in particular cases. How do we know that a particular person who got cancer got it *because* they smoked? Even if we know, as we do, that smoking causes cancer, it does not follow that a particular case of cancer was caused by smoking. Non-smokers also get cancer for other reasons, and not all smokers get cancer. Our question in this last section is how to move from generic causal knowledge to knowledge about particular cases.

The shift from generic to singular causes is typically marked by a change of tense. For instance, we say, generically, in the present tense, that smoking causes cancer, but we move to the past tense when we make claims about actual causes in a single case, saying for instance that Gabriel's smoking caused his cancer. This fact reminds us that finding actual causes is a matter of looking backward into the past, seeking events in the effect's past that are causally related to it.

Just how one should go about finding actual causes depends very much upon the kind of event you are seeking causes of, and success in this work depends greatly upon expertise. Epidemiologists are needed to find the cause of a disease outbreak; IT security specialists might look for the cause of a data breach; meteorologists might seek the cause of a sudden temperature drop. But while the methods used to investigate causes in these cases are diverse, at an abstract level, all seem to follow a common approach that philosophers call inference to the best explanation (IBE).

The basic idea of IBE is simple enough. Cast around for any plausible sets of circumstances that, if they were true, would offer a causal explanation of the event *e*. Choose from these the one that, given all of your generic causal knowledge and knowledge of the circumstances, is the likeliest. Suppose (true story) that the sticks of butter left on my counter disappeared. What caused this? The best explanation I have is that the dog ate them. He is a bad boy who frequently gets up on counters while people are distracted, and he has a known weakness for butter. Moreover, I found him hiding under the table where he typically takes his treasures, licking his greasy lips. There were of course other possible explanations. Maybe my neighbor sneaked in the back door and stole them, or, more likely, I simply misremembered, and I never got the butter out in the first place. But given all the evidence, blaming the dog seemed best.

IBE is a ubiquitous inferential strategy, both in ordinary life and in the sciences, but it is full of hazards. For one thing, the best explanation is not always a good explanation. Suppose we are jurors at a murder trial, and both the defense and prosecution have offered explanations of how the murder occurred. The defense story, let's imagine, supposes the murder was committed by an unknown intruder, while the prosecution blames the defendant, who is the victim's spouse. If you find yourself in a position where you don't see evidence for the defense story, but you are worried about certain holes in the prosecution story, you should not vote to convict. Even if the best explanation is the prosecution story, it's not enough to infer beyond a reasonable doubt that the defendant is guilty. The moral of such stories is that IBE should be modified to the principle that we should infer the best of the explanations which meet some minimal standard of goodness.

But this leaves us with the second question: what makes an explanation good, and by what principles should we judge one explanation better than another? One might be tempted to say that the best explanations are those most likely to be true, but that would be circular. For the purpose of identifying causal explanations of particular events though, we might reasonably say that the best explanation is the one that best accounts for known particular facts using known generic causal relationships and mechanisms.

Let's illustrate this approach by considering two cases of actual causation. Freddie Mercury, the lead singer for the band Queen, and the

American tennis player Arthur Ashe, both died of AIDS, but the actual causes of their becoming infected with HIV were different. What appears most likely is that Ashe was infected by a blood transfusion while Mercury was infected through sex with an HIV positive partner. How though do we know this?

We begin with the generic causal knowledge that the four major pathways for transmitting HIV/AIDS are unprotected sex, IV drug use with shared needles, blood transfusions, and transmission from pregnant mothers to unborn children. The mother to child path can be ruled out in both cases, since both Ashe and Mercury became infected in adulthood, and, at any rate, HIV was not circulating widely at the time they were born. Let's suppose also that neither of them were IV drug users. Arthur Ashe, however, did receive blood transfusions in a heart bypass surgery in 1983, during a time when blood supplies were not being effectively screened for HIV. Moreover, at that time Ashe's only sexual partner was his wife, who did not have HIV. The best explanation (which Ashe and his doctors believed) was that HIV must have been contracted during the transfusion. Mercury, on the other hand, did not receive blood transfusions, but he did have sexual partners in the gay community in London at a time when HIV infection was relatively common, so the best explanation seems to be sexual transmission.

It should be emphasized that what is the best explanation could change if new evidence were to arise. If say, we discovered that Ashe had been using IV drugs and sharing needles with an HIV positive community of addicts, we might revise our judgment. But, given what we know of Ashe, such a scenario seems farfetched, and we can comfortably accept that the actual cause of Ashe's infection was the transfusion.

We'll conclude our discussion of actual causation with a scientific case that is somewhat more complex: is CO_2 emission the principal cause of climate change? The answer to this question is, at this point, an unequivocal yes, though until the last decade or so, there was a good deal of political (though not scientific) resistance to accepting the answer (Oreskes and Conway 2010).

A great deal of evidence has accumulated supporting the fact that CO_2 emissions are the primary cause of global warming, but to simplify our discussion I want to focus on an argument made in a single influential paper that appeared in the journal *Science* (Hansen et al. 1981), that clearly exhibits an IBE structure. The particular event that Hansen and his collaborators attempted to explain was a 0.4°C increase in average global temperature between 1880 and 1980. Global warming skeptics pointed out that, while, overall, global temperature had increased during this time, most of the climb occurred between 1880 and1940, and that there had in fact been a period of cooling between 1945 and 1960, even though CO_2 levels continued to rise. The skeptics argued that the rise in global temperatures could be seen as a consequence of periodic variations in the amount of energy coming from the sun, and that a period of cooling would likely soon ensue.

To counter this argument, Hansen *et al.* built a model which included all three of the mechanisms hypothesized to be drivers of climate change. First, CO_2 emissions, via the Greenhouse effect, can trap heat within the atmosphere. Second, the earth can warm or cool in response to periodic variations in the amount of solar radiation. Third, aerosols in the stratosphere from volcanic eruptions will tend to lower temperature by reflecting the solar radiation away from the earth. Putting together the results of models of these three mechanisms, the scientists generated a graph of global temperature changes that very closely approximated the actual data. The component models showed that the overall warming was driven by CO_2 emissions, while smaller perturbations in the temperature were generated by solar and volcanic gas variations. In particular, the dip in global temperatures between 1945 and 1960 was caused by a series of volcanic eruptions which led to a brief period of global cooling.

Hansen *et al.*'s explanation was a good one, both because it closely matched the actual trajectory of climate change, and because it provided an explanation of the variations. The story has gotten even clearer in the forty years since, as the atmosphere has warmed by another 0.4°C, more than doubling the rate of increase from the previous century.

The tale of the doubt mongering surrounding the causes of global warming is a cautionary tale. While for more than fifty years there has been a scientific consensus that anthropogenic climate change is real and will have serious impacts, the complexity of the evidence was such that political actors were able to create an impression of uncertainty that will have lasting consequences.

SUGGESTIONS FOR FURTHER READING

A classic rejoinder to Hume's argument that causation cannot be observed is Ducasse's "On the Nature and Observability of the Causal Relationship" (1993). Beebee's "Causation and Observation" (2009) offers a contemporary defense of the observability of causal relationships, and Danks' "The Psychology of Causal Perception and Reasoning" (2009) offers a survey of relevant contemporary psychological research. Illari and Russo's *Causality: Philosophical Theory Meets Scientific Practice* (2013) is rare among philosophical introductions to causality in its emphasis on epistemological questions about causation, and it contains extensive discussions of scientific methods of observation, experiment and statistical inference. It also contains a chapter on causal inference in medicine that discusses both the Bradford Hill criteria and evidence-based medicine. Finally, I recommend Oreskes and Conway's *Merchants of Doubt* (2010) for a history of how in the last fifty years scientific communities have come to infer causation in complex cases like smoking and cancer, the effects of DDT, and global warming, and correspondingly of the ways in which political and economic actors have managed to contest or obscure scientific consensus about these causal claims.

10

CHOOSING CAUSES

Let me tell you what caused me to write this book. A couple of years back, the philosophy editor at Routledge contacted me to suggest I write a volume on causation for their Basics series. After we talked, I submitted a proposal, which his editorial board accepted, and here I am! This quick causal explanation of how I came to write this book is correct so far as it goes, but it is highly selective. So many other things can count of causes of my writing this book. Why choose this?

As we learned in Chapter 5, the lawyer's standard test for causation is the but-for test: *X* is an actual cause (or a cause in fact) of *Y*, if, but for the occurrence of *X*, *Y* would not have occurred. The events I cited in my quick explanation all meet this test—but for the editor contacting me, I wouldn't have written the book, etc.—but many other things also meet this test. The editor contacted me because I've written a good bit about causation, and but for my writings in this area, I'm sure he wouldn't have. And for all I know I was choice number two or three for this project, and but for prospective authors one and two turning him down, I wouldn't have been asked. We can go further of course. But for some disappointments in my first job out of college, I would not have decided to study philosophy and acquired the relevant expertise. More distantly, but for the various accidents of our parents' love lives, neither I nor my editor would have existed, so we would have been unable to strike the bargain. And but for the big bang, nothing would have happened.

DOI: 10.4324/9781003200956-10

Clearly, when we explain things, we select only a small fraction of these causes. The question of this chapter is how do we, and how should we, make these choices. One might hope that something in the world's causal structure could provide selection principles. Perhaps there are some objective facts about this structure that would allow us to distinguish causes from background conditions, or principle causes from secondary causes. However, most philosophers, myself included, think that causal selection is ultimately a practical matter. We choose the causes we do because those choices are helpful to us in some way, and what particular selections we make will depend upon the goals we have. But to say that these choices are practical is not to say either that they are arbitrary or unimportant.

Before turning to these principles, we must be clear on what causes we can select from. We have seen in previous chapters that there is some debate about what entities can count as causes or effects. For instance, we wondered whether the true relata of the causal relation are facts, or events, and whether we can count such things as agents, reasons, absences, or omissions as causes. Regardless of our metaphysical views though, as we enter the practical world of causal selection, it is appropriate to take a permissive approach, allowing all of these things to count as causes. It is possible to give different metaphysical accounts of what these things really are; but however we construe them, they can be important parts of the causal explanations we offer. Passing the but-for test, which is really just a counterfactual difference-making criterion, should be enough to allow something to be a target for causal selection. And we should also be liberal in allowing as causes goings on that don't pass the but-for test, but which for other reasons appear to be causes in fact. We would, for instance, want to count each of four bank robbers' actions as causes of a robbery, even though their individual actions fail the but-for test, since the heist would still have gone on even if one of them had not shown up for the job.

Since the purpose of selecting causes is to explain effects, I'll begin with an overview of the nature of explanation, and causal explanation in particular. After that I'll identify and illustrate some selection principles that can be used to decide which causes to include in explanations. Finally, I'll apply these principles to three actual cases, to see how we can use these principles to better understand and act in the world.

EXPLANATION: THE BASICS

Explanation is at its heart the activity of describing what depends upon what, and learning what depends upon what provides for us some measure of understanding. A tradition, which dates back to Aristotle, interprets explanations as answers to why questions (Van Fraassen 1977). Aristotle's four causes correspond to four different ways to answer a why question, reflecting four different kinds of dependence relations. The material cause shows what stuff a thing depends upon; the formal cause shows how a thing's nature depends upon its organization or form; the final cause shows for what purpose a thing exists; and the efficient cause, which corresponds most closely to the modern conception of cause, identifies agents, actions or events that brought the thing about.

Sometimes we speak of explanation not as an activity that we engage in, but as a relationship in the world itself. It is not uncommon to describe some facts as explaining other facts, as when we say things like "global warming is explained by carbon emission." When the word "explain" is used in this way, it is often just a synonym for "cause." In this chapter though I will use "explanation" to refer to something we do. *We* explain the earth's global warming by *showing* that it causally depends upon CO_2 emissions. More generally, philosophers understand explanations as showing that a phenomenon to be explained (called the **explanandum**) depends upon something else (called the **explanans**).

One of the advantages of thinking about explanations as answers to why questions is that it places the explanation in a concrete context. An explanation must be offered by someone, and it has an audience. What the exact question is and what would count as an adequate answer will depend upon the knowledge, interests and goals of the audience. Answering explanatory questions involves representing or describing parts of the world, and every representation will leave some things out, and gloss some things over.

Explanation is both a scientific activity and an activity of everyday life. In the personal sphere we look for explanations of why our partner is late home from work, what explains a friend's excitement or sadness, or why we did or didn't get a job. In the scientific sphere, we may try to explain why the weather in California is so severe, why the bee population is crashing, or why the wage gap between men and women persists. We seek to explain both singular things—events, actions, omissions, etc.—and general patterns or regularities.

Whenever one identifies some cause of an effect, one has at least the beginnings of an explanation, since effects depend upon their causes. What we might call a bare causal explanation is simply an explanation that identifies an event or kind of event upon which the explanandum causally depends. As a simple example, consider sex and pregnancy. Human beings have presumably understood since prehistory that pregnancy is an effect of sex between biological females and males. We understand in general that if a woman is pregnant, she must have had sex with a man (leaving aside modern inventions like IVF, etc.). A bare causal explanation might identify a particular man, or a particular sexual encounter, which led to the pregnancy.

Understanding that sex causes pregnancy is very different from understanding how sex causes pregnancy. To understand how, one needs a mechanistic explanation. While a bare causal explanation identifies one or more factors upon which the explanandum depends, a mechanistic explanation describes how things work. A mechanistic model will show how various entities and activities are organized so as to produce or underlie the explanandum. Often such explanations will span mechanistic levels. A mechanistic explanation of fertilization and implantation of embryos in humans could, for instance, take into account goings on at the level of whole bodies, organs, tissues, cells, and macromolecules.

Another form of explanation, akin to or perhaps a form of mechanistic explanation, is narrative explanation. In explanatory narratives, one identifies a sequence or set of sequences of activities and events that collectively bring about the explanandum. We tend to think of narratives in fiction, biography and history, as when Jane Austen offers a narrative explanation of how Miss Bennet and Mr. Darcy overcome their initial antipathy to find true love, or when Frederick Douglass in his autobiography, accounts for the events and opportunities that led him to escape from his enslavers. Narratives, however, play important roles in scientific explanation. Scientific narratives can be either singular, as when geologists, describe the history of the formation of the Hawaiian Islands through the activities of volcanoes, or generic, as when evolutionary biologists describe the sequence of events involved in the process of speciation.

In practice, there is no clear boundary between bare causal explanations and mechanistic or narrative explanations. As one moves beyond identifying just a single cause, to understanding the constellation of conditions that explain the effect, one begins to acquire mechanistic knowledge. In the case of pregnancy, even those who don't understand the cellular biology of pregnancy may well understand some things about the mechanism—for instance, that the man must ejaculate inside the woman, that the woman must be of an age where she can menstruate, and be at a certain point within her menstrual cycle; this level of understanding of what depends upon what might help a couple understand, for instance, how to promote or prevent pregnancy.

Explanation is often contrastive. For instance, one might be interested in explaining why a child is biologically male rather than female, in which case an explanation might pick out the sex-determining chromosomes. The factors picked out by an explanation will depend upon the contrast. If we want to explain why Lisa married Jenny, as opposed to their breaking up, the explanandum might include facts that account for their patching up their relationship. If we want to explain why they married, as opposed to continuing to live together unmarried, we might look to other factors like the fact that same-sex marriage recently became legal, that they thought that marriage made parenting easier, etc.

While the general point of explanation is to show dependence, some explanatory information may be acquired simply by finding that the explanandum does not depend upon something. A physician, for instance, might conduct tests to rule out certain causes of a patient's symptoms, and if she does not thereby figure out what the symptoms depend upon, she gains some understanding by knowing something they don't depend upon.

While all explanations describe things upon which the explanandum depends, some explanations cite as explanatory factors, not actions or events, but structures. A structural explanation describes some bits of the world's causal structure which constrain or enable causal processes. Structural explanations can explain why things happen without describing the particular process that produced their occurrence. As an example, consider how I would explain where I ended up on a ski-slope. I might account for my final position by describing the particular path I traveled down the mountain, identifying all of my activities, all the obstacles I avoided, etc. This would be a narrative/mechanistic explanation of how I got there. But more simply and perhaps more illuminatingly, I could simply explain how I ended up at the bottom by pointing out some obvious facts about the mountain's shape, the snow, my skis, and gravity. No matter what path I take down the mountain, I will, because of these constraints, end up at the bottom.

One of the virtues of structural explanations is they unify our knowledge. The structural explanation of why I end up at the bottom of the ski slope is very much the same explanation I can offer for why everybody else ends up at the bottom of the ski slope, and indeed, more generally why skiers ski to from top to bottom rather than bottom to top. Whether a bare causal, mechanistic or structural explanation will best serve depends upon what precisely the explanatory target is, and what the goals of the explanation are. If, for instance, I want to know why my friend went down the hill faster than I did, the explanation will likely have to include facts about our particular paths down the hill or our skills as skiers, but if I simply want to explain why we both ended up at a certain spot, a look at the mountain's geography might be enough.

Before continuing with the story, let me offer a couple of cautions about the picture of explanation I've sketched. First, my claim that explanation is fundamentally about dependence is not universally accepted. Other important theories hold either that explanation involves showing how phenomena are covered by laws (Hempel 1965) or fit within unifying explanatory patterns (Kitcher 1989). These theories, which comport broadly with Humean and other reductive approaches to causation, have fallen somewhat out of favor in recent years, as causal and mechanistic theories of explanation have become more prominent. Still, these theories provide important insights about scientific explanation that deserve more discussion than I can offer here.

Second, the view that explanation is about dependence does not imply that all explanation is causal. Most obviously, there are explanations in domains like mathematics, the law, and moral theory, which have little or nothing to do with the world's causal structure. Additionally, many philosophers think that there are non-causal explanations in the sciences, though the boundaries between causal and non-causal explanations are not always clear (Reutlinger and Saatsi 2018). In fact, what I'm calling structural explanations count as non-causal by some standards. I'll leave these questions aside though, so we can focus on the principles by which we construct causal explanations.

CAUSAL SELECTION PRINCIPLES

Explanation is always selective. To offer a causal explanation is to pick things out of the world's causal structure that allow us to understand why an effect occurs, and perhaps to do other things as well. In this section

I'll offer some principles for making these selections. If you are looking to explain what caused something, the principles will suggest factors to consider, and if you are looking to assess an explanation offered to you, they will help you notice the selections made and omitted, and consider whether the selections are apt for the purposes the explanation serves.

1. Start by selecting the effect. To get an explanation started, the first step is to select an effect as the explanandum. What is selected and how it is described will influence later steps in the construction of an explanation.

The first question to consider is whether the explanation sought is singular or generic. For instance, do you want to explain what caused Joe's heart attack, or what causes heart attacks in general? A second question is whether you are trying to explain why an event occurred at all, or if you are rather trying to explain some aspect of an event. You could explain why Jenny and Lisa had a wedding at all, or alternatively you could explain why their wedding occurred in Las Vegas. You can also seek explanations for aspects of generic events. You might, for instance, want to explain why weddings have gotten smaller over the last decade, or to explain why there are more outdoor weddings in Hawaii than Texas.

Sometimes an important part of selecting the effect is setting out what counts as the occurrence of an effect. One cannot, for instance, explain causes of eating disorders, or ways to prevent them, without some clinical criteria for deciding whether an individual has an eating disorder, and how severe it is. Developing these criteria (in medicine, but also in other areas) requires expertise, and often the criteria are at least in part conventional.

2. **Select the category of cause to do the explaining.** We've seen in our analysis of causal relata that there are a number of metaphysically distinct categories of entities that can be identified as causes, including events, actions, omissions, agents, properties, and structures. Suppose for instance that we are looking to explain an episode in which a man is killed by police. One kind of explanation simply seeks to identify who caused the death. Another might trace events leading up to the death—a phone call that alerted police, a traffic stop—as well as actions undertaken by people at the scene—handcuffing, beating, etc. Another might look at omissions, like the failure of another officer or bystander to intervene when an officer is using excessive force. Yet another might point to properties of the individuals involved in the incident, for instance a

police officer's history of reprimands, or the physical or emotional state of the victim. Finally, one might look at more structural causes—for instance, in police training, media representations, and implicit bias.

All of these kinds of causes can be useful components of explanations, but they serve different purposes. Event-based explanations are useful for instance in reconstructing the basic causal processes and interactions that led to an outcome, and also for asking certain counterfactual questions about what would have happened had certain events not occurred. Identifying agents as causes is useful in the practice of assigning praise or blame. Identifying structural causes may be helpful in understanding an effect as being part of a larger pattern of events, and identifying paths for systematic interventions.

3. **Select causes that violate norms**. If car *A* goes streaking through a red light and car *B* brakes suddenly but still hits the rear fender of car *A*, which car caused the accident? Clearly the motion of both cars is causally relevant to the accident's occurrence, but most people, including police officers and insurance adjusters, would judge car *A* to have caused the accident. The simple reason for this was that car *A*, by running the red light, did something that was both unusual and illegal. It violated a norm.

The terms "norm," "normal," and "abnormal" are used in overlapping ways that are important to untangle. In the quantitative or statistical sense of normality, what is normal is what is typical. We might say, for instance, that the rainfall in a year is normal, or that it is abnormally high or low. We also have a functional sense of normality, in which identifying a condition as abnormal means that it falls outside of a range of proper function. We might for instance identify a person's blood sugar levels as abnormally low, when they are at levels that could impact the body's functioning and even lead to death. Finally, norms can be understood as various kinds of standards—moral, legal, social, professional, etc.—with abnormalities being violations of these standards.

Norm violators in all of these senses are good targets for causal selection in circumstances where the explanandum is unusual or surprising. Suppose that one semester the average scores on my students' logic tests went up substantially from their typical levels. If, looking at my students, I discovered that an unusually large number of them were math majors, and that these students had scored highly, this departure from my normal mix of students would provide an informative causal explanation of the change. Similarly, we look for abnormalities in the functional sense when explaining failure. We might for instance cite as a cause of infection an immunodeficiency disorder that resulted in the infected person's body developing abnormally low concentrations of antibodies. Finally, as in the example of the car crash, it's reasonable to identify the violator of the traffic rules as the cause, for the purposes of assessing fault with respect to the accident.

Conversely, we typically will not find norm violators among causes when the explanatory target is some phenomenon that is regular and unsurprising. A causal explanation of things that happen all the time—regular motions of the tides, growth and development of cells or organisms, or non-malfunctioning behavior of artifacts like washing machines—will identify as causes things that themselves happen all the time.

Norm violation is particularly important in understanding when we should count an omission as a target for causal selection. Suppose my plants died while I was on vacation. I blame my neighbor because I'd asked him to water the plants, and he forgot. But while his failure meets the basic counterfactual test for causation—had he watered, the plants would have lived—that same counterfactual test is met by many other omissions. Had the mayor or the Pope watered my plants, my plants also would have lived. The reason I don't select the mayor's and the Pope's omissions as causes is that their omissions don't violate norms. My neighbor did, both in the sense that watering my plants is something he does regularly, and that it is something he promised to do.

4. **Select causes with larger effects.** When it is possible to quantify causal contributions to an effect, good explanations will select causes with larger effects. In some cases, where the effect is a quantity, we may be able to identify certain causes as having contributed more or less to the effect having the value it does. For instance, the degree to which intergenerational wealth transfer is appropriately cited as a cause of income inequality will depend upon how large a portion of a family's wealth arises from inheritance or gifts as opposed to savings from wages. In other cases, the size of the effect can be understood as an increase in the likelihood of an effect given the cause. Suppose that in some population drinking two drinks a day increased the risk of heart attacks by 20% over non-drinking, while eating red meat more than twice a week increased that risk by 2%. The drinking would then be a better cause to cite in an explanation.

The large effect principle I'm describing here should be distinguished from the statistical concept of effect size. Effect size in statistics is a measure of strength of association, and does not guarantee causation. Large effect size provides evidence for causation—it's Bradford Hill's first criterion—but it is possible for strong associations to occur without there being causation. If though a causal link is established, we should tend to select as explanatory those causes which make the biggest differences to their effects.

5. **Select causes that control effects**. Controllers are causes, but not all causes are controllers, and all things being equal, controllers provide more informative explanations of why events occur than other causes. If one is explaining the outcome of a battle, it is generally more informative to explain the activities of the general than the private. Without the privates there will be no battle, but the overall outcome is much less sensitive to the actions of a single private than it is to the general. There are exceptions to this principle; sometimes the best and most informative explanations of the battle will look at the soldiers (or the weather, or the weapons, and so on). Sometimes non-controlling factors provide good explanations because they meet other selection criteria, like norm violation. Control, as noted in Chapter 8, is a key aspect of agency, so identifying controlling causes will be helpful in explaining who or what is responsible for something, and who we should praise or blame.

6. **Select causes that unify**. Generally speaking, good causal selections will provide explanations that unify—explanations that not only account for what a particular event depends upon, but will show how they fit within a common pattern. Often the best explanations find the common thread that runs through what might at first seem disparate particular events. Structural explanations are generally unifying explanations, because persistent structures often enable or constrain repeated patterns of events or behaviors. For example, a structural explanation of police violence towards Black communities in the US might point to widespread and entrenched conditions, like police training, qualified immunity for police officers, and militarization of police forces. Selecting these factors, much more than exploring the misconduct of a single officer, can explain not just a particular incident, but a much wider pattern of misconduct.

7. **Select a level of causes that meets practical goals**. We encountered in Chapter 7 the idea that nature is organized into levels, and that causal mechanisms span levels. Consider for instance the different levels at which one can explain memory loss. Sometimes the best explanation of memory loss is neurological. Patients with Alzheimer's disease

lose their memory because it kills neurons and breaks down synaptic connections in various brain regions, including the hippocampus. This damage is due to a number of factors, but primary among them is excess concentration of amyloid proteins which clump together into plaques that disrupt cellular function. Other times the best explanation of memory loss is psychological. Dissociative amnesia is a diagnostic classification used by psychologists to characterize autobiographical memory loss not attributable to neurological conditions. Dissociative amnesia is often caused by trauma, as for instance when a victim of sexual assault cannot identify their attacker or remember details of the attack.

Of course, all psychological events have neurological bases. If one comes to believe something, or forget something, or to engage in some behavior, these effects depend upon neurological activities in the brain. Nonetheless there are good practical reasons to identify neurological causes for Alzheimer's and environmental and psychological causes for dissociative amnesia. The biggest reasons have to do with interventions. Clinical treatments for Alzheimer's are quite limited, but those that work to slow the progress of disease work by targeting amyloid build-ups, and if Alzheimer's is ever to be cured, it will likely be by medical interventions that stop the destruction of neurons and neural connections. Conversely, there are no drug therapies that target dissociative amnesia, and the most effective interventions involve various forms of talk therapies. Given though that dissociative amnesia is grounded in the brain, it is possible that neurological interventions may become possible in the future. For instance, research into the effects of psychedelics suggest that they may, in combination with talk therapy, may be helpful in addressing dissociative amnesia and other symptoms of trauma.

8. **Select not just single causes, but constellations of causes whose interactions make a difference to the effect**. While no explanation can possibly include all of the causes of an effect, that does not mean that selection principles should be used to identify a single or principal cause. There are some explanatory contexts when a single cause explanation is appropriate and justified, but more often, quality explanations will identify a constellation of causes, and will show how the interaction between these factors is responsible for the effect. Mechanistic and narrative explanations inevitably involve showing how a collection of causally relevant factors are organized so as to bring about an effect, and more complete accounts of how these factors are organized provide better understanding of the causes of the explanandum, and potentially

better understanding of how to intervene to meet goals. While in general, richer accounts of how these factors interact will improve an explanation, there are times when details are irrelevant or can interfere with the purposes of the explanation. More details are not always better.

CHOOSING CAUSES: SOME CHALLENGING CASES

To close our discussion of causal selection, I want to consider how the causal selection principles identified above can illuminate three cases where explanations have been highly contested, and where the things to be explained are of serious public consequence. The first concerns the causes of two fatal crashes of the 737 Max Airliner in 2018 and 2019. The second concerns the causes of the disparity in maternal death rates among White and Black American mothers. The third concerns the causes of the January 6 Capitol insurrection.

The 737 Max: On October 29, 2018, Lion Air Flight 610 took off from Jakarta in good weather conditions, crashing only 13 minutes later in the Java Sea. Within a month, the Indonesian National Transportation Safety Committee (NTSC) was able to issue a report on the causes of the crash. The problems began three minutes into the flight with the malfunctioning of an Angle of Attack (AOA) sensor, which detects how far the plane's nose is above or below the flight path. The malfunctioning sensor reported that the AOA was about 20 degrees higher than it in fact was, triggering something called the Maneuvering Characteristic Augmentation System (MCAS). The MCAS is control software designed to prevent the plane from stalling. When it detects an angle of attack close to the stall angle, it automatically pushes the nose down to bring it to a safer level. Because during this flight MCAS was relying on a reading from a malfunctioning sensor, the correction it ordered pushed the plane into a dive. The pilots tried to pull the plane out of the dive, but the MCAS repeatedly pushed the nose back down, leading ultimately to the fatal dive into the sea.

Thought of as a one-off event, an aircraft crash has countless causes—including all the causes that allowed the plane to take off in the first place. But in our explanatory practices, we understand aircraft crashes to be a specific kind of event—an accident. Investigations into accidents select as causes those systems, actions, or events that we can understood as malfunctions, failures, or mistakes. These investigations are both backward looking, figuring out who or what is to blame, and forward looking, figuring out what steps need to be taken to prevent similar accidents in the future.

Given the narrative I've sketched, one might think investigators would identify the malfunctioning AOA sensor as the primary cause of the crash, but they did not. The reason presumably is that sensors like this, which are attached to the plane's exterior, can become damaged in normal operating conditions. The failure instead was attributed to the design which linked control of MCAS to a single sensor. It is a design norm in engineering that any component whose failure could be catastrophic should have a backup so the failure of a single component will not lead to a catastrophic failure of the system as a whole.

Another source of failure was in aircraft maintenance. The AOA sensor on the plane had previously malfunctioned, and had recently been replaced, but the replacement had not been recalibrated. In fact, in its last flight before the fatal crash, the aircraft had experienced some of the same difficulties with sudden dives as occurred on Flight 610.

Initial reports also suggested that pilot errors may have been a contributing cause of the accident. There were in fact steps the pilots could have taken to disable the MCAS system and bring the aircraft back under control, but they failed to take these steps. The key question though is whether in the circumstances they had enough time and training to properly diagnose and respond to the problem. The NTSC report decided they did not, instead placing blame on cockpit design and training issues that made it difficult for pilots to respond to the malfunction.

The investigation was still ongoing when, only five months after the Lion Air crash, a second Boeing 737 Max, Ethiopian Airlines Flight 302, crashed six minutes after taking off from Addis Ababa. The circumstances were very similar, with the plane taking off in clear skies, reporting difficulties with flight control one minute after takeoff, and crashing six minutes later. As with Lion Air 610, flight control problems were caused by a malfunctioning AOA sensor triggering the MCAS, which began pushing the aircraft into a dive.

The occurrence of this second crash within five months, and within only two years of the first operational flights of the 737 Max, rapidly changed the narrative. Boeing had, after the first crash, insisted that the 737 Max was safe, and suggested that it could keep flying while it made some minor software fixes and augmented some training materials. After the second crash, authorities around the world grounded the airplane, and a series of investigations and redesign and certification efforts kept the plane from flying for nearly two years.

The second crash shifted the target of explanation. While the initial question concerned what caused a particular flight to crash, subsequent investigations sought to understand what caused an unsafe aircraft to be certified to fly. Why did Boeing engineers design the system as they did, and why did the FAA let them get away with it?

The brief story, which I draw largely from (Kennedy 2022) is this: The 737 Max is the fourth generation of the 737 —an airframe design dating from the 1960s. The main change from the third generation was the introduction of new engines that were larger and more fuel-efficient, which would allow the 737 Max to compete with a new generation of Airbus aircraft that were also more fuel-efficient. Engineers discovered that mounting the larger engine on the aircraft made the plane more prone to pitch up and stall, so they developed the MCAS to automatically push the nose down in such circumstances.

It was commercially advantageous to Boeing to get the plane quickly to market, and to minimize training requirements for pilots already trained on the 737. In particular Boeing wished to avoid requirements for flight simulator training, which would be costly and slow the rate at which 737 Maxes could be deployed. Boeing convinced the FAA that the differences between the 737 Max and earlier 737 models were minor, and they did not even mention the MCAS in training materials. For this reason, when the emergency occurred on Lion Air 610, the pilots did not even know what the MCAS was, much less what to do to turn it off. The pilots on Ethiopian 302 were only slightly better off. In response to complaints from pilots after the Lion Air crash, Boeing had provided some training information on MCAS, including information on how to turn the system off. The crew of Ethiopian 302 had in fact turned the system off, but that action also disabled automatic trim controls, which contributed to their inability to regain control of the plane.

Given all these facts, what shall we say caused these plane crashes? In the first instance the cause was the malfunction of the MCAS, and the lack of redundancy in sensors. This was the unifying explanation linking the two crashes. It's also the target of redesign. But the larger story is one of causation by omission—a series of omissions, chiefly by Boeing. During the rollout of the 737 Max, Boeing failed to notify the FAA, pilots and airlines of the scope of changes to the plane's design. In particular, they did not notify them of the existence and purpose of the MCAS or train them how to respond to malfunction, even though internal safety assessments at Boeing had identified MCAS malfunction as a point of potential catastrophic failure. There was also regulatory failure, because the FAA exercised insufficient oversight. These failures are apt choices of causes because they are violations of norms—professional, legal, and moral. They also tell us what agents to blame—chiefly Boeing and its executive leadership, but also the FAA. These were the agents in control of the decisions that put a dangerous aircraft in the air. We do not blame the pilots, since even if there were actions they could have taken that would have prevented the crashes, given the pilots' knowledge and training it was not reasonable to expect that they would be able to do so.

Stepping back further, one can find structural causes for these norm violations. The deepest of these appears to be the degradation of safety culture at Boeing, as a consequence of shareholder capitalism. Boeing executives' decision to push a new plane quickly to market and to cut corners on training were driven by efforts to retain or increase market share and to increase corporate profits. These are structural in the sense that Boeing, like other publicly traded corporations, is judged by its profits and stock price. While it is clearly reasonable to hold particular executives to account for their decisions, it is far from clear given these structural conditions that other executives would have done that much better.

Black maternal mortality in the United States: Understanding the causes of maternal mortality presents a rather different kind of challenge than understanding the causes of the 737 Max plane crashes. Childhood maternal mortality (defined by the World Health Organization (WHO) as deaths of mothers during pregnancy or childbirth, or within 42 days of birth for any cause related to or aggravated by pregnancy) is typically measured by deaths per 100,000 births. A 2023 report from the Centers for Disease Control and Prevention (CDC) paints a disturbing picture of maternal mortality in the United States. Some key data points are indicated in the following table (Hoyert 2023, "All women" includes women of other races, mixed races, or races not specified):

TABLE 10.1 US maternal mortality rates per 100,000 births

	2018	*2019*	*2020*	*2021*
All women	17	20	24	33
Non-Hispanic Black	37	44	55	70
Non-Hispanic White	15	18	19	27
Hispanic	12	13	18	28

Maternal Mortality Rates in the United States are the worst among major industrialized nations (though roughly tied with China). WHO data for 2020 reports an infant mortality rate for the US that is seven times that of Australia and Spain, five times that of Germany and Japan, and twice the rate of the UK and Canada. The problem is much worse in some other countries. Nigeria for instance is among the worst, with a rate of over 1000 deaths per 100,000 births. But the rate in the US is shocking, given its overall wealth and healthcare spending.

One of the principles of explanation we've articulated is that explanations often seek to account for contrasts, and there are several contrasts we might consider. We might for instance wish to explain why the US mortality rate has risen from 2018 to 2021, or why US mortality rates are so much higher than mortality rates of other major industrialized nations. These different explananda will require different choices of causes. For the purpose of this example, let's focus on the fact that Black American women consistently die at more than twice the rate of American women overall. The explanandum in this case is not a particular event—e.g., when a particular woman dies as a result of pregnancy or childbirth—but a generic fact about a disparity within the population of American women.

Pregnant or postpartum women can die from a number of complications in pregnancy and childbirth. The most common causes include hemorrhaging, hypertensive disorders, infections, complications of unsafe abortions, and complications, owing to the interaction between pregnancy or birth and preexisting conditions like heart disease. Most of these conditions are not fatal given timely medical intervention, so it is likely that the explanation of why some mothers die while others do not will largely concern quality of and access to healthcare.

One immediate question to consider is whether race as such directly affects maternal mortality rates, or whether those effects are mediated by other variables. For instance, Black woman have on average less income and wealth than White women, so it could be that the income and wealth gap explain the racial disparity. It turns out, however, that the disparity persists, even when controlled for income and wealth. Highly educated Black women also die at rates far greater than White women of comparable education.

One of the most commonly identified explanations of the disparity between Black and White American women is the fact that Black mothers are more likely to have preexisting conditions which put them at risk during pregnancy. Rates of heart disease, hypertension, diabetes, and obesity are all higher among Black mothers than non-Black mothers. A related explanation is based upon a phenomenon called weathering. Geronimus (1992) found evidence that the cumulative effects of socio-economic disadvantage caused Black mothers to exhibit health profiles and outcomes comparable to older White mothers, so that, for instance, the risks associated with a 35 year old Black woman would be comparable to that of a 42 year old White woman. Subsequent research has found molecular markers of aging, showing that the shortening of DNA segments called telomeres, a biological effect of aging, occurs faster in Black women than in White women (Geronimus et al. 2010). The accelerated biological aging is thought to be due to chronic stress caused by a variety of circumstances, from overt racism and exposure to violence, to stresses induced by not enough sleep, overly long commutes, housing insecurities, and all of the other effects of systemic racism.

An important question about selecting causes is how far one looks back from the effect. In pointing out that Black mothers have higher rates of obesity, heart disease or hypertension as risks for maternal mortality, one has in a sense explained at least part of the healthcare disparity. But these differing rates of preexisting conditions themselves have causes rooted in systemic racism. These factors, for instance, depend upon access to healthy foods, which may be difficult for many Black mothers to obtain, both because they are living in food deserts and because healthier foods are often more expensive than less healthy foods. And the cost differentials in these foods are in turn largely explained by American agricultural and industrial policies that subsidize highly processed high calorie foods.

The second major explanation of the disparity is differences in access to quality healthcare. Black mothers, for instance, have on average less access to prenatal and postnatal care than White mothers. Much but not all of these differences are in turn explained by socio-economic disparities, but even controlling for these, there are race-specific effects. One well-documented phenomenon that affects a variety of health outcomes is implicit bias (Hall et al. 2015). Healthcare professionals have been shown, for instance, to spend less time with patients of color than White patients, and to take less seriously reports of symptoms.

Reflecting on how best to explain a complex population-level phenomenon like maternal mortality disparities provides a number of insights into causal selection. First it shows the value of a multi-level approach. To understand and attack the problem it is important both to identify the population-level phenomena and its high-level causes, and to understand the more immediate biological and social mechanisms that impact individual pregnancies. Second, it is important to select causes that violate norms. Many of the norms in this case are moral or professional. Healthcare, most would argue, is a human right, and differential access to healthcare based on race or other demographic factors is an injustice. Third, it is helpful to identify causes that unify. High Black maternal mortality rates are part of a broader pattern of health and healthcare disparities that affect all Black people, as well as other persons of color, the poor, and so on. One reason that attending to such unifying causes is important is that policy interventions to address them would address a wider range of health disparities.

Causal selection should be attached to the possibility of intervention, and for this reason, different actors should make different selections. The medical community is rightly concerned with implicit bias as a cause of healthcare disparities, because such causes have the potential to be corrected through better medical training. From a government policy perspective, more attention should be directed to differential healthcare access as a cause, as various policies, from Medicaid expansion to changes in zoning, could reduce these differences.

Causal selection should also be sensitive to the fact that any causes at a population level are averages across different causal structures in different subpopulations. The causes that explain maternal mortality rates in Nigeria are very different than those that explain them in the US, and within the US, the causes will be different in Mississippi than Massachusetts. This implies that explanations and interventions should be tailored to these populations.

The Capitol insurrection: The storming of the US Capitol on January 6, 2021 will undoubtedly go down as an important moment in American history, and already a great deal of investigation and analysis has been undertaken to understand its causes. Evidence has been collected and presented in President Trump's impeachment trial, in congressional oversight hearings, and in the work of the House Select Committee to Investigate the January 6th Attack on the United States Capitol. It has also been the subject of numerous investigations by the United States Department of Justice, some of which have resulted in convictions of participants and some of which, as of this writing, remain ongoing. Given that the basic facts of this event are well known, I shall not rehearse them. I want rather to use our selection principles to explore the basis for identifying particular agents, events or omissions as the causes, or the principal causes, of the insurrection.

To start, we should consider why it is not explanatory to cite as a cause of the insurrection the fact that Trump lost the election. Trump's loss passes the but-for test, since there would have been no violent attempt to storm the Capitol to stop the certification of an election that Trump had won, but we do not pick out Trump's loss as a cause, because the loss itself is not a norm-violation. Whenever there is a contested election, someone loses, but it is a norm of well-functioning democracies that losers concede their losses, and that incumbents who have lost assist in the peaceful transfer of power to the winner. It is not Trump's loss, but the norm-violating actions of Trump and other agents after the loss, which explain the insurrection.

The first step in causal selection is, we've seen, selecting the target, and in calling the happenings of January 6 the Capitol insurrection, we have already made a choice. In the period since the event, experts and journalists struggled to say what kind of event happened that January 6. Was it a protest, a riot, an insurrection, or a failed coup? The choice matters for deciding what has to be explained. By choosing to classify it as an event of a certain kind, one is making a choice about what was essential to the occurrence of the event. If it was just a riot, for instance, it could have happened even had it not disrupted the proceedings at the US Capitol. The choice also has implications for how we assess praise or blame for the event, and how seriously we take the need for remedies. It is not surprising that many politicians and media figures implicated in the events leading up to January 6 have claimed that the event was not an insurrection.

The second selection decision is to choose between an explanatory focus on agents, events or structures. All three focuses are of value in developing a complete understanding of the causes of the insurrection, but I will focus here on agents and structures. Agents are causes to the extent that their actions, aspects of their actions, or failures of action are causally relevant to the target. Agent causal explanations are valuable because they identify who or what is responsible, and relatedly, who should be held to account.

The actions (or inactions) of many agents meet the minimal standard of causal relevance, so the selection question is to identify which one or which few to single out as the causes of the insurrection. Let's consider a few of the most obvious candidates, starting with President Trump.

Much of the work of the House Select Committee focused on identifying Trump as the principal cause of the insurrection. As Representative Liz Cheney memorably put it, Trump "summoned the mob, assembled the mob and lit the flame of this attack." Trump is an obvious choice of cause for several reasons. First, many of his actions satisfy the but-for test. At numerous points between the election and January 6, had Trump not acted (or failed to act) as he did, the insurrection would not have occurred. Second, and relatedly, Trump was in control. He was, as Cheney's words suggest, the general controlling the mob's actions. Third, his actions were unprecedented in the history of US presidential elections, and they violated a host of constitutional obligations and democratic norms.

One might blame those members of the mob who actually entered the Capitol building as causes, and indeed many of them have been held criminally liable for their actions—but choosing these agents as causes is much less informative than choosing Trump and other leaders. For one thing, the actions of individual insurrectionists don't meet the but-for test. While they participated in the event, the event did not depend for its occurrence on any one or few of the people who actually broke in. Better selections would identify agents, like the heads of the paramilitary groups, who planned the Capitol attack, because it is likely that, but for their actions, the insurrection would not have occurred. Again, choose as causes agents who are in positions of control.

Some early congressional hearings focused on identifying failures by the leaders of intelligence and security agencies as causes of the event. Choosing these agents (e.g., the Senate Sergeant of Arms, the head of the Capitol police) as causes of the insurrection probably is not warranted. Their actions, or inactions, may have been blameworthy, but those actions arguably are more central to explaining aspects of how the event played out rather than the fact the event occurred at all.

While it is natural to select as causes agents who actually planned and participated in the insurrection, there are many others whose actions or inactions are enough to warrant selecting them as causes. A prominent collection of agents who are causally implicated in the insurrection are many leaders within the Republican party. The insurrection would not have occurred had not a large number of followers of President Trump been convinced that the election was stolen. Even if Trump claimed the election was stolen, had a significant majority of congressional Republicans spoken up loudly and clearly after the election to congratulate President-elect Biden and to denounce as disinformation unsubstantiated claims of election fraud, it is unlikely that the insurrection would have occurred. Similarly, had not numerous political officials at state and local levels pursued strategies to contest election results, there would have been no story of injustice to motivate the mob.

Another set of agents who can be selected as causes are the news organizations and media figures who failed to challenge or who even promoted false claims about election fraud. Many of those who would organize or participate in the insurrection were influenced by the misinformation provided by these outlets and personalities. These choices are also made more salient by the norm violations involved. Journalistic norms demand that news organizations should try very hard to get the facts right. They should avoid promoting false or unsubstantiated claims, and when newsworthy figures make such claims, they should clearly mark them as false or unsubstantiated. We would not identify careful and truthful reporting of Biden's victory as a cause of the insurrection, even if seeing that reporting helped motivate insurrectionists, but we do highlight the false reporting as a cause, because it is norm-violating.

While identifying the agents is a valuable part of explaining the causes of the insurrection, many have argued that there are deeper structural causes that are ultimately more important. Let me mention three that seem salient. First, one could argue that the insurrection was caused by the electoral college system. This system has numerous features that contributed to the conditions for the insurrection, and that, in general, make disputed elections more likely. Most immediately, it is only because of the electoral college system that the US Congress had to meet to certify electors, so without it, the event that was the target of the insurrection would not have occurred. But the deeper difficulty is that the state-by-state winner-take-all approach to selecting electors makes election outcomes depend upon potentially narrow margins of victory in a small number of battleground states. In the 2020 election, there was no question that Biden won the popular vote, but claims of fraud in a few tightly contested battleground states was enough, in the minds of many Trump supporters, to throw the election into doubt.

A second set of structural factors concern the media environment. Over the past fifty years, Americans' sources of news and political commentary have become increasingly balkanized. Whereas in the 1970s, most people got their news from the three major news networks, the emergence of cable news and talk radio in the 1980s and 1990s led networks to target audiences with particular ideological leanings. This balkanization has allowed different citizens to live in different worlds. The problem has only been exacerbated by algorithms used by social media organizations like Facebook and Twitter, which feed people news they want to hear, and which can easily spread misinformation. Given this media environment, it is not surprising that commentators at Fox News and other conservative media outlets were disinclined to share information that would have turned off their viewers.

A third cause, perhaps the deepest structural cause of the insurrection is the systemic racism that pervades American history and institutions. Those who point to racism as a root cause of the Capitol insurrection see the events of January 6 as in keeping with a history of White people using violence to resist the authority of democratically elected governments. They point, for instance, to parallels to other acts of political violence, like the Wilmington insurrection of 1898, where a group of white supremacists successfully forced the mayor of Wilmington and other members of the municipal government (both Black and White) from office. They also point out that political institutions, from the electoral college to the construction of the current two-party system, reflect legacies of racism.

The choice between focusing on the particular agents and events, or looking for underlying structural causes depends upon explanatory goals. Agent and event causal explanations inevitably focus on understanding the causal structures leading to a particular event, while structural causes show how the event was made more likely by different structural affordances. Structural causal explanations also serve the explanatory aim of unification, because the persistence of structural causes can be used to account for larger patterns of events. This makes structural causes potentially more forward looking, since modifying these structures may provide the best intervention for preventing future occurrences of an unwanted pattern.

FINAL THOUGHTS

It would take far more than I've said here to provide a satisfactory analysis of the causes of the 737 Max crashes, high Black maternal mortality rates, or the Capitol insurrection. My purpose in this discussion was only to take these familiar and important cases to illustrate some of the challenges of causal selection, and the principles that can be used to meet them. Looking back at these cases, we can draw several important lessons.

First, causal explanations involve choices, and it is both possible and useful to give multiple legitimate explanations of the same effect. Relatedly, just because one agent or event can be selected as a cause, care must be taken not to assume it is the only or most important cause. Sometimes causal explanations point to some factors while ignoring others, in ways that can obscure the larger causal structure. Third, what causes one should select depends upon what one is in a position to change—and because different observers of an event have different abilities and roles, different things will be appropriate to identify as causes. In the case of the 737 Max crashes, the engineers should identify different causes than the pilots or the airline executives. In the case of maternal mortality, med schools should choose different causes than insurance companies or legislators. In the case of the Capitol insurrection, politicians, activists, security officials, members of the news media, and voters should all choose different causes, because all of them have different roles in making the event happen, or causing or preventing like events in the future.

When someone selects a cause or offers a causal explanation, they do so with a purpose. In pointing at the causes they do, they identify what to worry about, who to praise or blame, or what to do. Sometimes the purposes are nefarious, as when politicians, businesses, or interest groups point the finger at some cause that allows them to shift the blame. At other times, the selection is well intentioned but may reflect a parochial standpoint that ignores important factors and fails to see the larger patterns. It is my hope that making explicit what can or should go into causal selection will make you a more educated consumer and producer of causal claims. Causes, and only causes, can change the world, so we should learn to choose well.

SUGGESTIONS FOR FURTHER READING

Salmon's *Four Decades of Scientific Explanation* (1990) provides an excellent survey of the development of theories of explanation, and the particular challenges of causal explanation. Ross and Woodward's (2023) *Stanford Encyclopedia* article provides a survey of recent developments, and Chapter 7 of *The New Mechanical Philosophy* (2017) offers my own take on varieties of causal and non-causal explanation.

Hitchcock and Knobe's "Cause and Norm" (2009) discusses the role of norms as selection principles, while Ross's (2018)" Causal Selection and the Pathway Concept" emphasizes the role of control. A more general overview of the problem of selecting and integrating explanations of complex phenomena, especially in the context of policy, is Mitchell's (2009) *Unsimple Truths*.

GLOSSARY

Axiom: A statement in a mathematical or scientific theory that is assumed to be true and from which other true statements can be derived.

Background Conditions: Facts about the world or a part of it that generally hold true and that enable or restrict relations between causes and effects.

Causal model: A representation of causal structure, often but not always represented using structural equations or directed acyclic graphs.

Cause in fact: In legal settings, any fact or event that is causally relevant to the occurrence of an effect, regardless of whether it is legally significant.

Cause, efficient: Aristotle's term for an action or event that produces change. The kind of Aristotelian cause that corresponds most closely to contemporary concepts of cause.

Cause, final: Aristotle's term for the reason or purpose for a thing.

Cause, generic: A cause that often or always produces or makes a difference to an effect within a population, or in a kind of individual.

Cause, proximate: In the law, a cause in fact that is primary or is foreseeable and for which a defendant can be held civilly or criminally liable.

Cause, singular: A particular event, action, fact etc. that produces or makes a difference to a single effect at a particular place and time.

Cause, structural: A set of conditions that enables or constrains particular causal processes and relations. Related to background conditions.

Ceteris paribus: All things being equal.

Conditional: An if-then statement. The if clause is called the antecedent and the then clause is called the consequent.

Conditional, counterfactual: A conditional in which the antecedent condition is false.

Conditional, material: A conditional that is interpreted truth functionally, such that it is true so long as either the antecedent is false or the consequent is true.

Constitution: The relation that holds between a thing (object, process, event, etc.) and its parts. Constitution is often contrasted with causation.

Convention: Facts or principles are said to be conventional when their truth or correctness is not fixed by nature, but instead depend in whole or in part on practical, political, social or legal decisions and norms.

Determinism: The view that the complete state of the world at a time, together with the laws of nature, fixes all future states of the world.

Determinism, hard: The view that no will is truly free, because determinism is true, and that determinism is incompatible with genuine freedom of the will.

Determinism, soft: The view that determinism is true, but that under certain conditions free will may still exist.

Directed Acyclic Graph (DAG): A kind of model used to represent causal structure consisting of a set of variables (the nodes) and unidirectional arrows connecting them (the edges).

Difference making: Causes are said to make a difference to their effects if the occurrence of a cause is necessary for, or sufficient for, or makes more likely an effect, or if the effect depends counterfactually on the cause.

Disconnection: A causal scenario in which an effect occurs because a connection which was inhibiting or preventing the effect is blocked (disconnected). Also called double prevention.

Dualism: The view that minds are non-physical substances or have non-physical properties.

Epistemology: The theory of knowledge.

Explanans and explanandum: In an explanation, the explanandum is the event, fact or circumstance that is to be explained, while the explanans is the collections of events, facts or circumstances that do the explaining.

Generalization: A statement describing some general pattern or regularity.

Generalization, universal: A generalization, like "All humans are mortal," that holds always and without exception.

Generalization, existential: A generalization, like "Some birds fly," that implies that something is true of one or more individuals.

Generalization, *ceteris paribus*: A generalization, like "When prices rise, demand falls," that holds true unless some unspecified range of typically occurring background conditions fails to hold.

Law of Nature: Patterns or regularities that are universal, necessary or non-accidental; also, generalizations or principles describing these patterns.

Libertarianism: In metaphysics, the view that genuine free will exists and is incompatible with determinism, implying that determinism is false.

Mechanism: Systems or processes consisting of a collection of entities acting and interacting in ways that produce or underlie some phenomenon.

Metaphysics: The study of the most basic constituents of and relations between things – for instance, matter and mind, space and time, causation, determinism and freedom of the will.

Modality: The modality of a statement depends upon whether it asserts something as necessary, actual, possible, impossible, etc.

Necessary Condition: To say that x is a necessary condition for y is to say that, if y occurs or is true, then x will occur or be true.

Necessity: A statement is necessary if it must be true, an event is necessary if it must occur, and so on. Necessity comes in many varieties, including logical, metaphysical, causal, legal, and moral.

Norm: A social, political, cultural or legal principle or standard that guides action. Also a pattern of activity that is typical.

Omission: A causal scenario in which an effect occurs because an action or event which would have prevented the effect fails to occur.

Ontology: The study of what, in the most basic sense, there is.

Overdetermination: A causal scenario in which two or more causes are each sufficient for an effect.

Physicalism: The view that everything that exists is ultimately physical.

Preemption: A causal scenario where there is an action or event A that will cause an effect E, but where another action or event B occurs and causes E before A can do so.

Prevention: A causal scenario where an effect would occur, but in fact does not occur, because a preventative action or event blocks or inhibits events or processes that would cause the effect.

Probability, subjective & objective: Subjective probabilities measure degrees of belief or estimates of likelihood, while objective probabilities measure the frequency of or propensity of outcomes within physical systems.

Production: Causes are said to produce effects if there is some force or process by which the cause is connected to and brings about the effect.

Reduction: An analysis that seeks to define one thing (the reduced) in terms of another (the reducer), thereby showing that what is reduced is ultimately nothing but what it is reduced to. For instance, a reductive analysis might claim that causation is nothing but constant conjunction, or genes are nothing but DNA.

Regularity, accidental vs. non-accidental: A regularity is accidental to the extent it arises due to chance and is not determined by causes or laws.

Regularity, causal vs. non-causal: Causal regularities are regular patterns of associations between causes and effects. While causal regularities are non-accidental, some non-accidental regularities are not causal, as for instance when there is a regular association of two effects of a common cause.

Relation: In logic, relations are connections of various kinds between ordered pairs or larger ordered sets of individuals, for instance, relations between cause and effect, or between earlier and later times, or between parent and child. The individuals connected by relations are called their relata.

Screening off: Variable B screens of a variable A from a variable E if conditionalizing on B eliminates any probabilistic dependence of E on A, i.e., if $P(E \mid B \& A) = P(E \mid B)$.

Semantics: The study of the meaning, reference and truth conditions for words, phrases, and sentences.

Sufficient condition: To say that x is a sufficient condition for y is to say that, if x occurs or is true, then y will occur or be true.

Universal: In logic and philosophy, universals are properties or relations that can be instantiated or exemplified by multiple entities.

REFERENCES

Allen, Colin and Jacob Neal. 2020. "Teleological Notions in Biology." In *The Stanford Encyclopedia of Philosophy*, edited by Edward N. Zalta, Spring 2020. Metaphysics Research Lab, Stanford University. https://plato.stanford.edu/archives/spr2020/entrieseleology-biology.

Anscombe, G E M. 1971. *Causality and Determination: An Inaugural Lecture*. Cambridge: Cambridge University Press.

Armstrong, David M. 1983. *What Is a Law of Nature?* Cambridge: Cambridge University Press.

———. 1993. "The Identification Problem and the Inference Problem." *Philosophy and Phenomenological Research* 53 (2): 421–22. https://doi.org/10.2307/2107779.

Beebee, Helen. 2009. "Causation and Observation." In *The Oxford Handbook of Causation*, edited by Helen Beebee, Christopher Hitchcock, and Peter Menzies, 471–97. Oxford: Oxford University Press.

Beebee, Helen, Christopher Hitchcock, and Peter Menzies. 2009. *The Oxford Handbook of Causation*. Oxford: Oxford University Press.

Bradford Hill, Austin. 1965. "Causation and Association." *Proceedings of the Royal Society of Medicine* 58 (5): 295–300. https://doi.org/10.1016/S0079-7421(08)60562-9.

Cartwright, Nancy. 1979. "Casual Laws and Effective Strategies." *Nous* 13 (November): 419–37.

Collins, John, Ned Hall, and L. A. Paul, eds. 2004. *Causation and Counterfactuals*. Cambridge, MA: MIT Press.

Craver, Carl F. 2007. *Explaining the Brain*. Oxford: Oxford University Press.

Craver, Carl F and James G Tabery. 2016. "Mechanisms in Science." In *Stanford Encyclopedia of Philosophy*, edited by Edward N. Zalta. http://plato.stanford.edu/archives/win2015/entries/science-mechanisms.

Danks, David. 2009. "The Psychology of Causal Perception and Reasoning." In *The Oxford Handbook of Causation*, edited by Helen Beebee, Christopher Hitchcock, and Peter Menzies. Oxford: Oxford University Press. https://doi.org/10.1093/oxfordhb/9780199279739.003.0022.

Dennett, Daniel C. 1984. *Elbow Room*. Cambridge, MA: Bradford Book/MIT Press.

———. 1989. *The Intentional Stance*. Cambridge, MA: MIT Press.

Descartes, Rene. 1984. "Meditations on First Philosophy." In *The Philosophical Writings of Descartes, Vol. II*, translated by John Cottingham, Robert Stoothoff, and Dugald Murdoch, 1–62. Cambridge: Cambridge University Press.

Dowe, Phil. 1992. "Wesley Salmon's Process Theory of Causality and the Conserved Quantity Theory." *Philosophy of Science* 59 (2): 195–216.

———. 2000. *Physical Causation*. Cambridge: Cambridge University Press.

Dretske, Fred. 1977. "Laws of Nature." *Philosophy of Science* 44 (2): 248–68.

Ducasse, Curt J. 1993. "On the Nature and Observability of the Causal Relation." In *Causation*, edited by Ernest Sosa and Michael Tooley, 125–36. Oxford: Oxford University Press.

Ehring, Douglas. 2009. "Causal Relata." In *Oxford Handbook of Causation*, edited by Helen Beebee, Christopher Hitchcock, and Peter Menzies, 388–313. Oxford: Oxford University Press.

Frankfurt, Harry. 1971. "Freedom of the Will and the Concept of a Person." *The Journal of Philosophy* 68 (1): 5–20.

Geronimus, Arline. T. 1992. "The Weathering Hypothesis and the Health of African-American Women and Infants: Evidence and Speculations." *Ethnicity & Disease* 2 (3): 207–21.

Geronimus, Arline T., Margaret T. Hicken, Jay A. Pearson, Sarah J. Seashols, Kelly L. Brown, and Tracey Dawson Cruz. 2010. "Do US Black Women Experience Stress-Related Accelerated Biological Aging?: A Novel Theory and First Population-Based Test of Black-White Differences in Telomere Length." *Human Nature* 21 (1): 19–38.

Glennan, Stuart S. 2017. *The New Mechanical Philosophy*. Oxford: Oxford University Press.

Glennan, Stuart S, Phyllis Illari, and Erik Weber. 2022. "Six Theses on Mechanisms and Mechanistic Science." *Journal for General Philosophy of Science* 53:143–161

Hájek, Alan. 2019. "Interpretations of Probability." In *The Stanford Encyclopedia of Philosophy*, edited by Edward N. Zalta, Fall 2019. Metaphysics Research Lab, Stanford University. https://plato.stanford.edu/archives/fall2019/entries/probability-interpret.

Hall, William J., Mimi V. Chapman, Kent M. Lee, Yesenia M. Merino, Tainayah W. Thomas, B. Keith Payne, Eugenia Eng, Steven H. Day, and Tamera

Coyne-Beasley. 2015. "Implicit Racial/Ethnic Bias Among Health Care Professionals and Its Influence on Health Care Outcomes: A Systematic Review." *American Journal of Public Health* 105 (12): e60.

Hansen, J., D. Johnson, A. Lacis, S. Lebedeff, P. Lee, D. Rind, and G. Russell. 1981. "Climate Impact of Increasing Atmospheric Carbon Dioxide." *Science* 213 (4511).

Heil, John. 1998. *Philosophy of Mind: A Contemporary Introduction*. London: Routledge.

Hempel, Carl G. 1965. *Aspects of Scientific Explanation, and Other Essays in the Philosophy of Science*. New York: Free Press.

Hitchcock, Christopher. 1995. "Discussion: Salmon on Explanatory Relevance." *Philosophy of Science* 62 (2): 304–20.

———. 2007. "Three Concepts of Causation." *Philosophy Compass* 2 (3): 508–16.

———. 2021. "Probabilistic Causation." In *The Stanford Encyclopedia of Philosophy*, edited by Edward N. Zalta, Spring 2021. Metaphysics Research Lab, Stanford University. https://plato.stanford.edu/archives/spr2021/entries/causation-probabilistic.

———. 2023. "Causal Models." In *The Stanford Encyclopedia of Philosophy*, edited by Edward N. Zalta and Uri Nodelman, Spring 2023. Metaphysics Research Lab, Stanford University. https://plato.stanford.edu/archives/spr2023/entries/causal-models.

Hitchcock, Christopher and Joshua Knobe. 2009. "Cause and Norm." *Journal of Philosophy106* 106 (11): 587–612.

Hoyert, Donna. 2023. "Maternal Mortality Rates in the United States, 2021." National Center for Health Statistics (U.S.). https://doi.org/10.15620/cdc:124678.

Hume, David. 1975. *Enquiries Concerning Understanding and Concerning the Principles of Morals*. Edited by Selby-Bigge, L.A. and Nidditch, P.H. 3rd ed. Oxford: Clarendon Press.

———. 1998. *Dialogues Concerning Natural Religion*. Indianapolis: Hackett Publishing. https://www.amazon.com/Dialogues-Concerning-Natural-Religion-Classics/dp/0872204022/ref=sr_1_1?crid=3GLB6TEDI3YR7&keywords=hume+dialogues+concerning+natural+religion+cambridge&qid=1686333565&s=books&sprefix=hume+dialog%2Cstripbooks%2C84&sr=1-1.

Illari, Phyllis McKay, and Federica Russo. 2013. *Causality: Philosophical Theory Meets Scientific Practice*. Oxford: Clarendon Press.

Kane, Robert. 2005. *A Contemporary Introduction to Free Will*. New York: Oxford University Press.

Kant, Immanuel. 1965. *Critique of Pure Reason*. Translated by Kemp Smith, Norman. New York: St. Martin's.

Kennedy, Rory, dir. 2022. *Downfall: The Case Against Boeing*. Netflix. https://www.netflix.com/search?q=downfall&jbv=81272421.

Kim, Jaegwon. 1989. "The Myth of Nonreductive Materialism." *Proceedings and Addresses of the American Philosophical Association* 63 (3): 31–47.

———. 1998. *Mind in a Physical World: An Essay on the Mind-Body Problem and Mental Causation*. Cambridge, MA: MIT Press.

Kitcher, Philip. 1989. "Explanatory Unification and the Causal Structure of the World." In *Scientific Explanation (Minnesota Studies in the Philosophy of Science, Volume XI)*. Minneapolis: University of Minnesota Press.

Kuhlmann, Meinard, and Stuart S Glennan. 2014. "On the Relation between Quantum Mechanical and Neo-Mechanistic Ontologies and Explanatory Strategies." *European Journal for Philosophy of Science* 4 (3): 337–59.

Lewis, David K. 1973a. "Causation." *The Journal of Philosophy* 70: 556–67.

———. 1973b. *Counterfactuals*. Oxford: Blackwell.

Mackie, J. L. 1965. "Causes and Conditions." *American Philosophical Quarterly* 2 (4): 245–64.

———. 1980. *The Cement of the Universe: A Study of Causation*. Oxford: Clarendon Press.

Marks, Gary N. 2008. "Accounting for the Gender Gaps in Student Performance in Reading and Mathematics: Evidence from 31 Countries." *Oxford Review of Education* 34 (1): 89–109.

Mayr, Ernst. 1961. "Cause and Effect in Biology." *Science* 134 (November): 1501–6.

Mill, John Stuart. 1874. *A System of Logic, Ratiocinative and Inductive: Being a Connected View of the Principles of Evidence and the Methods of Scientific Investigation*. 8th ed. New York: Harper and Brothers.

Mitchell, Sandra D. 2009. *Unsimple Truths: Science, Complexity and Policy*. Chicago: University of Chicago Press.

Moore, Michael. 2019. "Causation in the Law." In *The Stanford Encyclopedia of Philosophy*, edited by Edward N. Zalta, Winter 2019. Metaphysics Research Lab, Stanford University. https://plato.stanford.edu/archives/win2019/entries/causation-law.

Mumford, Stephen. 2009. "Causal Powers and Capacities." In *The Oxford Handbook of Causation*, edited by Helen Beebee, Christopher Hitchcock, and Peter Menzies, 265–78. Oxford: Oxford University Press.

Nagel, Thomas. 2012. *Mortal Questions*. Canto Classics. Cambridge: Cambridge University Press. https://doi.org/10.1017/CBO9781107341050.

Newton, Isaac. 1999. *The Principia: The Authoritative Translation and Guide: Mathematical Principles of Natural Philosophy*. Translated by I. Bernard Cohen, Anne Whitman, and Julia Budenz. University of California Press.

Oreskes, Naomi., and Erik M. Conway. 2010. *Merchants of Doubt: How a Handful of Scientists Obscured the Truth on Issues from Tobacco Smoke to Global Warming*. New York: Bloomsbury Press.

Paul, L. A. 2004. "Aspect Causation." *The Journal of Philosophy* 97 (4): 235–56.

Paul, L. A. and Ned Hall. 2013. *Causation: A User's Guide*. Oxford: Oxford University Press.

Paul, Sarah K. 2022. *Philosophy of Action: A Contemporary Introduction*. New York: Routledge.

Pearl, Judea. 2009. *Causality: Models, Reasoning, and Inference*. 2nd ed. Cambridge: Cambridge University Press.

Povich, Mark and Carl F Craver. 2018. "Mechanistic Levels, Reduction, and Emergence." In *The Routledge Handbook of Mechanisms and Mechanical Philosophy*, edited by Stuart S Glennan and Phyllis McKay Illari, 185–97. London: Routledge.

Psillos, Stathis. 2009. "Regularity Theories." In *Oxford Handbook of Causation*, edited by Helen Beebee, Christopher Hitchcock, and Peter Menzies, 131–57. Oxford: Oxford University Press.

Reardon, Sean F., Erin M. Fahle, Demetra Kalogrides, Anne Podolsky, and Rosalía C. Zárate. 2018. "Gender Achievement Gaps in U.S. School Districts (CEPA Working Paper No. 18.13)." http://cepa.stanford.edu/wp18-13.

Reichenbach, Hans. 1956. *The Direction of Time*. Berkeley, Calif.: University of California Press.

Reutlinger, Alexander and Juha Saatsi. 2018. *Explanation Beyond Causation: Philosophical Perspectives on Non-Causal Explanations*. Edited by Alexander Reutlinger and Juha Saatsi. Oxford: Oxford University Press.

Ross, Lauren N. 2018. "Causal Selection and the Pathway Concept." *Philosophy of Science* 85 (4): 551–72. https://doi.org/ru.

Ross, Lauren and James Woodward. 2023. "Causal Approaches to Scientific Explanation." In *The Stanford Encyclopedia of Philosophy*, edited by Edward N. Zalta and Uri Nodelman, Spring 2023. Metaphysics Research Lab, Stanford University. https://plato.stanford.edu/archives/spr2023/entries/causal-explanation-science.

Russell, Bertrand. 1913. "On the Notion of Cause." *Proceedings of the Aristotelian Society* 13: 1–26.

Salmon, Wesley C. 1980. "Probabilistic Causality." *Pacific Philosophical Quarterly* 61: 50–74.

———. 1981. "Causality: Production and Propagation." In *PSA 1980: Proceedings of the Biennial Meeting of the Philosophy of Science Association*, 2: 49–69.

———. 1984. *Scientific Explanation and the Causal Structure of the World*. Princeton: Princeton University Press.

———. 1990. *Four Decades of Scientific Explanation*. Minneapolis: University of Minnesota Press.

———. 1997. "Causality and Explanation: A Reply to Two Critiques." *Philosophy of Science* 64: 461–77.

Schaffer, Jonathan. 2000. "Causation by Disconnection." *Philosophy of Science* 67 (2): 285–300.

———. 2005. "Contrastive Causation." *Philosophical Review* 114 (3): 327–58.

Sellars, Wilfred. 1963. "Philosophy and the Scientific Image of Man." In *Empiricism and the Philosophy of Mind*, 1–40. London: Routledge & Kegan Paul.

Stalnaker, Robert. 1976. "Possible Worlds." *Noûs* 10 (1): 65–75.

Strawson, Galen. 2014. *The Secret Connexion: Causation, Realism, and David Hume: Revised Edition*. Revised edition. Oxford: Oxford University Press.

Van Fraassen, Bas C. 1977. "The Pragmatics of Explanation." *American Philosophical Quarterly* 14 (2): 143–50.

Williamson, Jon. 2009. "Probabilistic Theories of Causality." In *The Oxford Handbook of Causation*, edited by Helen Beebee, Peter Menzies, and Christopher Hitchcock, 185–212. Oxford University Press.

———. 2011a. "Mechanistic Theories of Causality Part I." *Philosophy Compass* 6: 421–32.

———. 2011b. "Mechanistic Theories of Causality Part II." *Philosophy Compass* 6: 433–44.

Woodward, James. 2003. *Making Things Happen*. Oxford: Oxford University Press.

INDEX